T0305824

The Mathematical Foundations of Mixing

Mixing processes occur in a variety of technological and natural applications, with length and time scales ranging from the very small – as in microfluidic applications – to the very large – for example mixing in the Earth's oceans and atmosphere. The diversity of problems can give rise to a diversity of approaches. Are there concepts that are central to all of them? Are there tools that allow for prediction and quantification?

The authors show how a range of flows in very different settings – micro to macro, fluids to solids – possess the characteristic of *streamline crossing*, a central kinematic feature of 'good mixing'. This notion can be placed on firm mathematical footing via Linked Twist Maps (LTMs), which is the central organizing principle of this book.

The authors discuss the definition and construction of LTMs, provide examples of specific mixers that can be analysed in the LTM framework and introduce a number of mathematical techniques – nonuniform hyperbolicity and smooth ergodic theory – which are then brought to bear on the problem of fluid mixing. In a final chapter, they argue that the analysis of linked twist maps opens the door to a plethora of new investigations, both from the point of view of basic mathematics as well as new applications, and present a number of open problems and new directions. Consequently, this book will be of interest to a broad spectrum of readers, from pure and applied mathematicians, to engineers, physicists, and geophysicists.

CAMBRIDGE MONOGRAPHS ON
APPLIED AND COMPUTATIONAL MATHEMATICS

Series Editors
M. J. ABLOWITZ, S. H. DAVIS, E. J. HINCH, A. ISERLES, J. OCKENDON,
P. J. OLVER

22   The Mathematical Foundations of Mixing

The *Cambridge Monographs on Applied and Computational Mathematics* reflects the crucial role of mathematical and computational techniques in contemporary science. The series publishes expositions on all aspects of applicable and numerical mathematics, with an emphasis on new developments in this fast-moving area of research.

    State-of-the-art methods and algorithms as well as modern mathematical descriptions of physical and mechanical ideas are presented in a manner suited to graduate research students and professionals alike. Sound pedagogical presentation is a prerequisite. It is intended that books in the series will serve to inform a new generation of researchers.

Also in this series:

The Mathematical Foundations of Mixing
The Linked Twist Map as a Paradigm in Applications Micro to Macro, Fluids to Solids

ROB STURMAN

School of Mathematics
University of Bristol
University Walk
Bristol, BS8 1TW
United Kingdom

JULIO M. OTTINO

Departments of Chemical and Biological Engineering
and Mechanical Engineering
Northwestern University
Evanston, Illinois 60208
USA

STEPHEN WIGGINS

School of Mathematics
University of Bristol
University Walk
Bristol, BS8 1TW
United Kingdom

CAMBRIDGE
UNIVERSITY PRESS

Shaftesbury Road, Cambridge CB2 8EA, United Kingdom

One Liberty Plaza, 20th Floor, New York, NY 10006, USA

477 Williamstown Road, Port Melbourne, VIC 3207, Australia

314–321, 3rd Floor, Plot 3, Splendor Forum, Jasola District Centre, New Delhi – 110025, India

103 Penang Road, #05–06/07, Visioncrest Commercial, Singapore 238467

Cambridge University Press is part of Cambridge University Press & Assessment,
a department of the University of Cambridge.

We share the University's mission to contribute to society through the pursuit of
education, learning and research at the highest international levels of excellence.

www.cambridge.org
Information on this title: www.cambridge.org/9780521868136

First published 2006

A catalogue record for this publication is available from the British Library

ISBN   978-0-521-86813-6   Hardback

THOMASINA:
When you stir your rice pudding, Septimus, the spoonful of jam spreads itself round making red trails like the picture of a meteor in my astronomical atlas. But if you stir backward, the jam will not come together again. Indeed, the pudding does not notice and continues to turn pink just as before. Do you think this odd?

SEPTIMUS:
No.

THOMASINA:
Well, I do. You cannot stir things apart.

SEPTIMUS:
No more you can, time must needs run backward, and since it will not, we must stir our way onward mixing as we go, disorder out of disorder into disorder until pink is complete, unchanging and unchangeable, and we are done with it for ever.

Arcadia, Tom Stoppard

Contents

Preface

Where is mixing important?

Mixing processes occur in a variety of technological and natural applications, with length and time scales ranging from the very small (as in microfluidic applications), to the very large (mixing in the Earth's oceans and atmosphere). The spectrum is quite broad; the ratio of the contributions of inertial forces (dominant in the realm of the very large) to viscous forces (dominant on the side of the very small) spans more than twenty orders of magnitude.

Theoretical and experimental developments over the last two decades have provided a strong foundation for the subject, yet much remains to be done. Earlier work focused on mixing of liquids and considerable advances have been made. The basic theory can be extended in many directions and the picture has been augmented in various ways. One strand of the expansion has been an incursion into new applications such as oceanography, geophysics and applications to the design of new mixing devices, as in microfluidics. A second strand is incursion into new types of physical situations, such as mixing of dry granular systems and liquid granular systems (in which air is replaced by a liquid). These applications clearly put us on a different plane – new physics – since, in contrast to mixing of liquids, a complicating factor in the flow of granular material is the tendency for materials to segregate or demix as a result of differences in particle properties, such as density, size, or shape. Mixing competes with segregation: mixtures of particles with varying size (S-systems) or varying density (D-systems) often segregate leading to what, on first viewing, appear to be baffling results. This class of problems can also be attacked with extensions of the basic theory.

The diversity of problems can give rise to a diversity of approaches and a temptation to deepen work on application-driven tools. Specificity may dominate the picture. One could, however, take the opposite approach. Are there concepts that are central enough that they should be developed in more detail? Broadness competes with specificity. Would however this viewpoint allow for a more encompassing view that may otherwise be lost by generation of fragmented results?

Why is this book needed?

The purpose of this book is to focus on new developments in mathematics and take this broader view point – the objective is on general results rather than specifics. Even though the work is of a manifestly mathematical bent, we expect that the presentation will resonate with a diverse set of readers. We are aware however that the question 'Is all this mathematics really necessary?' will surface in many readers' minds. The reason for the mathematics is to set a baseline, a clear picture as to where things come from. We recognize however, that in long stretches of the presentation things get manifestly technical, and we are also aware that we have to persuade the reader that going through the material is a good investment of time. We have therefore decided to provide help in navigating the book. Periodically we insert physical and heuristic explanations to go along with the mathematical descriptions. All chapters have mini-intros, a distillation of the contents of the chapters, and connections to fluid applications where possible.

What will this book cover?

It is also important to clarify what we will and will not do. Of all the components of mixing we will focus on only one – the kinematical aspects of mixing a fluid with itself. The objective here is to study a problem which is tractable from an analytical as well as a computational perspective. The broad objective, then, is to determine what characteristics of a flow enable it to efficiently stretch material lines and surfaces, and to analyse the simplest possible flows capable of 'good mixing'. Making precise what we mean by 'good mixing' will occupy us for at least one chapter of the book. We shall focus primarily on kinematics, rather than on dynamics (in the sense of the dynamics of fluids). Our study is an analysis of the motion due to an imposed velocity field; i.e. the study of the following dynamical system:

$$\mathrm{d}x/\mathrm{d}t = v(x, t), \quad x = X \quad \text{at} \quad t = 0.$$

In general, we shall study the motion by a sort of minimal flow – the simplest non-trivial flow that encapsulates the characteristics of wide classes of flows – and expose the characteristics which make them good or poor mixing flows, rather than study the hydrodynamic forces which give rise to the flows themselves. The goal is to make recent dynamical systems theory accessible to users interested in the mixing of fluids.

It is then clear that *our presentation is a purposefully distorted view of mixing*. There is nothing in this view, for example, that accounts for turbulent flows. We argue however, that the baseline presented here is crucial to the understanding of more complicated cases of mixing.

At the starting point of any mixing process there exist two (or more) constituents which occupy distinct domains whose size is on the order of the system size itself. The objective of the mixing process is to reduce the length scales of these materials below a certain level, resulting in a 'homogeneous' system – a mixture. This length scale and the level of homogeneity are of course dependent upon the application in question. We distinguish three aspects. The first is essential; the other two may or may not be present. (1) *Mechanical mixing*: This is the stretching and folding part; the motion causes the interfaces between the materials to stretch, creating inter-material area between highly striated structures. The system, which at first contained a blob of one fluid in another, now appears as a stretched and folded filament with, in general, a wide distribution of striation thicknesses; (2) *Breakup*: If the filament has been sufficiently stretched, differences in the interfacial tension on opposite sides of an interface can cause the filament to break up into isolated droplets, reducing the length scale even further; in this step the properties of the materials matter; in the case of very viscous materials or in the case of very similar materials this aspect may be absent; (3) *Diffusion*: If the fluids are miscible, Brownian motion of individual fluid molecules, due to fluctuations in thermal energy, acts to homogenize the fluid at the molecular scale. This process does not take place if the materials are incompatible.

Thus not all of these mechanisms need be present in a given mixing process and, in many cases, breakup may be accompanied by its opposite – coalescence or unmixing. As a specific example, consider the mixing of two polymeric fluids with similar, but large, viscosities. Diffusion coefficients in such systems can be on the order of 10^{-8} cm^2/sec. In this case, both breakup and diffusion are negligible, and the only means for mixing is stretching. In any case, the time required for molecular homogenization, T, can be estimated from simple dimensional analysis to be $T = L^2/D$, where D is the molecular diffusion coefficient and L is the length scale of the fluid domains. The most important factor which affects the time scale for final homogenization is the length scale of the fluid domains. The length scale is in turn determined by the extent of the stretching of the material domains which occurs due to the imposed motion. Thus, the most important part of understanding mixing is to understand how, and what types of flows are able, to generate efficient stretching.

How are the chapters related to one another?

The book is organized around the idea of the linked twist map (LTM). LTMs provide a minimal picture of mixing and one case for which nearly everything that can be known is known; for example, theorems from ergodic theory and conditions that guarantee mixing behaviour, measures of complexity, and geometrical properties. Results generally apply to infinite-time limits and one could argue that reality does not correspond to this case. As with every theory care is needed in interpreting its applicability. For example, a system having a horseshoe does not imply that good mixing will take place (by 'good' we invoke the notion of 'widespread' as opposed to 'localized'). On the other hand the absence of a horseshoe guarantees that no good mixing will occur. Most of the material presented is scattered throughout the mathematics literature. In particular we draw on four papers from the pure dynamical systems community, Devaney (1978), Burton & Easton (1980), Wojtkowski (1980), and Przytycki (1983) that were published around 25 years ago, mostly in arguably obscure and hard to find conference proceedings. Moreover their intended audience was a very different one from ours. Thus one of our main tasks, rather than to present a host of original dynamical systems theorems, is to distil the papers into the first unified and user-friendly presentation of these ideas and to show how these results and the mathematical details within the now classical proofs relate to contemporary mixing problems. The credit for the original pure mathematical results rests with the five authors listed above. In Chapters 1 and 2 we consider applications from a variety of fields; microfluidics, granular mixing as produced by tumbling, and transport in geophysical flows, for example, but we emphasize the 'universality' of the linked twist map approach to mixing across disciplines.

The book starts in full in Chapter 3, entitled 'The ergodic hierarchy'. This chapter is necessary because smooth ergodic theory is a technical subject requiring a mathematical background beyond that of most physicists, chemists, and engineers, yet we believe it is poised to play an important role in the subject of mixing in applications in the future. The situation is similar to that which existed for dynamical systems theory in the late 1970s and early 1980s. At that time the subject was relatively mathematically abstract and it required substantial extra effort on the part of physicists, chemists, and engineers to carry out the transference between 'abstract theorem' to a technique that could unlock the secrets of the specific nonlinear dynamical systems arising in applications. In smooth ergodic theory such a transference is only now beginning, but it promises to be equally as fruitful. Consider a specific example of the type of issue that techniques in smooth ergodic theory may address. The Smale horsehoe is the classic 'chaotic dynamical system', but it is only a set of 'zero area', and one would therefore expect that the probability of 'observing' the dynamics on this set would be zero. Experiments suggest otherwise – that the chaotic dynamics may persist beyond a set of zero area. Dynamical systems theory without smooth ergodic theory gives no indication that this might be the case. Smooth ergodic theory provides a framework for making these notions mathematically precise, and provides the techniques for extending results to sets of observational significance. In particular, the notion of the area of a set is dealt with quantitatively through the notion of measure and measurable sets. These notions allow one to give a probabilistic description of the dynamics, and also to quantify the idea of 'observability' in a way that lends itself to computation. Smooth ergodic theory also provides a framework for analysing a much more practical question – 'When does Smale horseshoe-like behaviour occur on a set of non-zero area?', or equivalently, 'when is the horseshoe observable?' By now Smale horseshoes have been shown to exist in literally hundreds of dynamical systems spanning many diverse areas of applications. However, their influence on 'observable dynamics' is unclear at best. One could view the linked twist maps studied in this monograph as being an example of a dynamical system exhibiting 'horseshoe-like' behaviour on a set of 'full measure', and ergodic theory provides the tools for carrying out the necessary analysis that quantifies this statement. But *mixing* is the motivating application for our studies, and ergodic theory enables us to characterize the mixing process mathematically and rigorously in a variety of ways. For example, it provides an ordered list of behaviours of increasing complexity; from ergodicity, through (measure-theoretic) mixing, to the Bernoulli property. We describe the main features of ergodicity, mixing, and the Bernoulli property in detail, as these are the most immediately applicable to the problem of fluid mixing. It should be noted that these definitions are very technical and differences between them may not be realizable in applications.

Chapter 4, 'Existence of a horseshoe for the linked twist map', builds on the definition and construction of a linked twist map on the plane. The underlying structure of complicated behaviour that arises in LTMs is that of the Smale horseshoe. This chapter presents a detailed construction of the horseshoe, and the implications of its existence for symbolic dynamics. The central element is revealing the existence of the invariant set of the LTM by a carefully chosen pair of quadrilaterals in the intersection of the two linked annuli. The invariant set is given by the images of the intersection of the quadrilaterals under infinite forward and backward time. The way in which the image of each quadrilateral intersects the original pair of quadrilaterals defines a *symbolic dynamics*, which gives a measure of the complexity of the system.

The horseshoe constructed is 'classical' in the sense that the chaotic invariant set is uniformly hyperbolic having measure zero. In this sense the linked twist maps provide a concrete example showing how nonuniform hyperbolicity arises in this class of dynamical systems. In later chapters we show how it can be analysed.

Chapter 5, 'Hyperbolicity', deals with one of the most fundamental aspects of dynamical systems theory, both from the point of view of pure dynamical systems – it represents one of the best-understood classes of dynamical system; and from the point of view of applied dynamical systems – one of the simplest models of complex and chaotic dynamics. We define uniform and nonuniform hyperbolicity, and go on to describe Pesin theory, which establishes a bridge between nonuniform hyperbolicity and the ergodic hierarchy. The theory of Pesin is central to our analysis. It relates non-zero Lyapunov exponents and nonuniform hyperbolicity. The method of invariant cone fields is introduced to prove the existence of non-zero Lyapunov exponents on a set of full measure. The special structure of the linked twist maps renders such calculations feasible. Once nonuniform hyperbolicity is established the theory of Pesin paves the way for conclusions about the existence of partitions into ergodic components and the Bernoulli property.

Chapter 6 is entitled 'The ergodic partition for toral linked twist maps' and discusses the application of Pesin theory in detail to linked twist maps defined on a torus. Here, drawing on three key papers from the ergodic theory literature, we give the proof that linked twist maps on the torus can be decomposed into (at most a countable number of) ergodic components.

Chapter 7 centres on 'Ergodicity and the Bernoulli property for toral linked twist maps'. Here we apply a global geometric argument, again from the ergodic theory literature, to extend the result of Chapter 6 to ergodicity and the Bernoulli property on a set of full measure for toral linked twist maps. Conditions are given such that these results hold. We give sufficient conditions for a toral linked twist map to enjoy the Bernoulli property. A key point – of significant practical importance – to notice is that different conditions are required for the co-rotating and counter-rotating cases.

In Chapter 8, these results are extended to planar linked twist maps. Linked twist maps on the plane are more directly applicable to fluid mixing, but introduce new technical difficulties in the mathematics.

Finally, in Chapter 9, we discuss a number of open problems. The analysis of linked twist maps only opens the door to a plethora of new investigations, both from the point of view of basic mathematics, as well as new applications. In fact, the latter drives the former as applications naturally suggest new types of linked twist maps, which we describe in detail. The sufficient conditions that we derive in earlier chapters leading to ergodicity and the Bernoulli property can be used as design parameters for optimizing mixing regions, and we show how this can be done.

In closing we remark again that in some sense this book can be read at two levels. Chapters 3 through 8 stand on their own as an analysis of the dynamics of linked twist maps on the torus and on the the plane, containing all of the necessary background and details. However, we believe that the real value of this approach comes in when one considers that linked twist maps embody the mixing paradigm of 'crossing of streamlines'. When the range of examples showing this in Chapters 1 and 2 is considered then it becomes apparent that one has a new way of looking at the mixing process that leads to characterizing mixing properties on large regions of the domain in a way that has not been done before. The fact that the approach feeds immediately into optimization and design makes it even more compelling.

Acknowledgments

RS and SW gratefully acknowledge the financial support of the Office of Naval Research and the Engineering and Physical Sciences Research Council.

JMO wishes to acknowledge the financial support of the John Simon Guggenheim Memorial Foundation.

We thank James Springham for a careful reading of the text.

RS would like to thank Kathryn for much patience and support.

1

Mixing: physical issues

This chapter provides a brief review of physical considerations in the analysis of mixing problems and several examples of problems that can be framed in terms of the mathematical structure covered in this book.

Mixing is a common phenomenon in everyday life. A blob of white cream placed in a cup of black coffee and gently stirred with a spoon forms, if one looks carefully, intricately shaped striated structures, until the mixture of coffee and cream homogenizes into a fluid that is uniformly brown in colour. This common phenomenon serves to illustrate some of the key features of mixing; namely, the interplay between advection and diffusion. If the coffee is at rest when the cream is added (and assuming that the insertion of the cream into the coffee only causes negligible disturbance of the surrounding coffee) then, in the absence of stirring, the cream mixes with the coffee by the mechanism of molecular diffusion. Experience tells us that in this particular situation the mixing takes much longer than we would typically be willing to wait. Therefore we stir the admixture of coffee and cream with a spoon, and observe it to homogenize very quickly. This stirring illustrates the role that advection plays in homogenizing the cream and coffee. In fact, in this particular example (as well as many others) the role of molecular diffusion in achieving the desired final mixed state may very well be negligible.

In this monograph we will concentrate exclusively on mixing via convective motions or advection. This is the foundation upon which the entire subject of mixing is built. Of course, the impact or lack thereof of molecular diffusion on mixing is a fact that requires justification, and this justification occurs within the physical context of specific mixing problems. The spectrum of problems occurring in nature and technology where mixing is important is enormously wide (see Figure 1.1). For example, in the subject of mantle convection (Kellogg (1993)) it probably seems reasonable that diffusion has essentially no impact on the mixing of 'rock with rock'. At the other end of the spectrum, in the realm of the very small, mixing in microfluidic devices is another area in which diffusion may have a negligible effect. In this setting the goals are to mix quickly and in

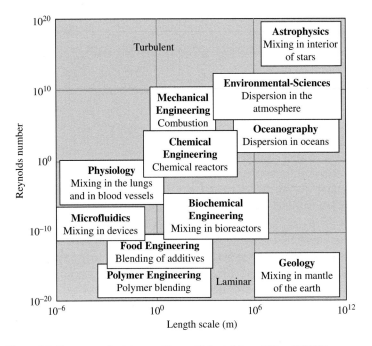

Figure 1.1 Spectrum of mixing problems. [Adapted from Ottino (1990).]

small spaces, and achieving these goals tends to make the effects of diffusion negligible and to prohibit the creation of turbulent flows, which are well known to enhance mixing. In fact the subject of mixing at the microscale is tailor made for the mathematical approach of 'chaotic mixing' and the dynamical systems approach, about which now there is a very large literature (see, for example Ottino (1989a, 1990), Wiggins (1992), Wiggins & Ottino (2004)).

The dynamical systems approach to mixing, in the absence of diffusion, is the central theme of this book. But more precisely, we develop the notion of the *linked twist map* (LTM) as a paradigm for chaotic mixing in that it embodies the kinematic mechanism of 'streamline crossing' as a mechanism for generating chaotic fluid particle trajectories. But most importantly, the LTM framework provides a way in which mixing can be optimized in the sense that one can give conditions under which mathematically rigorous characterizations of strong mixing occur on regions of nonzero area. Of course, the conditions leading to strong mixing in regions of nonzero area do not guarantee fast mixing, something one wishes to produce in practice. However, not satisfying the conditions guarantees that mixing will not be widespread, an outcome which is clearly undesirable. Thus, in a strict sense, the conditions described in this book are necessary conditions for effective mixing.

Before developing this approach in some detail, and describing how LTMs naturally arise in the context of a variety of mixing problems, we first consider some general physical and kinematic considerations of mixing in general that will provide rough, but essential, guides to understanding the issues relating to 'good' and 'bad' mixing.

1.1 Length and time scales

In any 'mixing problem' a consideration of length and time scales is fundamental as they provide an indication of the main mechanisms at work. Dimensional quantities, such as length and time scales, often combine with certain material parameters (e.g., molecular diffusivity, viscosity, etc.) to form dimensionless ratios that provide rough guides to the relative importance of competing mechanisms. The Reynolds number, Re, is the ratio of inertial forces to viscous forces. If U and L denote characteristic velocity and length scales, Re is UL/ν, where ν is the kinematical viscosity, which is the ratio of viscosity, μ, and density, ρ, i.e, $\nu = \mu/\rho$. Small values of Re correspond to viscous dominated (or laminar) flows, and large values of the Reynolds number correspond to turbulent flows (see examples in Figure 1.1). The Péclet number, Pe, is the ratio of transport by advection (or convection) and by molecular diffusion; Pe is defined as Pe $= UL/D$, where D is the molecular diffusion coefficient. Pe can be interpreted also as the ratio of diffusional to advective time-scales; the time scale for diffusion is L^2/D and the time scale for convection is L/U. A large value of Pe indicates that advection dominates diffusion, and a small Pe indicates that diffusion dominates advection, or, in terms of time-scales, the fastest process dominates. The ratio Re/Pe is ν/D, the ratio between two transport coefficients, the so-called Schmidt number, $Sc = \nu/D$. Sc can be interpreted as the ratio of two speeds. The speed of propagation of concentration is $\delta_D \sim (Dt)^{1/2}$, the speed at which concentration gets smoothed out, whereas the propagation of momentum is $\delta_V \sim (\nu t)^{1/2}$, the speed that it takes for motion to spread out or die. The ratio of these two speeds, $(d\delta_V/dt)/(d\delta_D/dt)$ is $Sc^{1/2}$; thus if $Sc \gg 1$, as in the case of liquids, concentration fluctuations survive without being erased by mechanical mixing until late in the process. We will encounter these and other numbers in the following examples. As a reference point the kinematic viscosity of water is about $0.01\,\text{cm}^2/\text{s}$ and of air $0.15\,\text{cm}^2/\text{s}$; somewhat surprisingly momentum spreads more quickly in air than in water. The value of ν in liquids is highly dependent on temperature. The diffusion coefficient of small molecules in water is about $10^{-5}\,\text{cm}^2/\text{s}$; thus a typical value of Sc for a liquid such as water is about 10^3. For gases Sc is of order one.

Example: mixing in a coffee cup

Consider again the case of mixing of milk in a coffee cup. Assume that the cup's characteristic length is $L \sim 4$ cm and that the typical speed is $U \sim 5$ cm/s. Then the Reynolds number is approximately 2,000, indicating that advection is much more important than viscous effects; a few strategic turns of the spoon get the job done. Even if the spoon is held in place the wake behind the spoon mixes the fluid (the wake flow behind a stationary object being a well-studied problem). Mixing of milk in golden syrup is another matter. The kinematical viscosity of golden syrup at 15°C is 1200 cm^2/s, so Re $\sim 10^{-2}$. In this case viscous effects dominate and one cannot rely on inertia; the spoon is removed and the motion stops. An estimate of the time it takes for the motion to die off is L^2/ν. In the case of syrup the motion stops in a hundredth of a second whereas in the case of milk the estimate is half an hour. Advection dominates molecular diffusion in both problems, Pe $\sim 10^6$ in the case of milk and syrup. The time necessary for mixing relying solely on molecular diffusion is L^2/D. The estimate in this case is in the order of more than a day for either problem.

Example: flow in a small channel

Consider the flow of two adjacent streams of fluid in a channel of length L along the z-direction having a cross-sectional area in the plane xy with a characteristic length h describing the width of the channel in the cross-section. The velocity in the z-direction is denoted $v_z(x, y)$ with a mean value U. In microfluidic applications typical numbers are $h \sim 200\,\mu$m, and $\mu/\rho \sim 10^{-2}$ cm^2/s. Take U as 1 cm/s. The Reynolds number in this case is Re $= Uh/\nu \sim 2$. This small value of the Reynolds number implies that flows in microfluidic channels are typically viscous dominated. The no-slip boundary condition on the walls of the channel leads to velocity profiles having parabolic shapes (i.e. at a given cross-section, $v_z(x, y)$ is zero on the walls, and increases monotonically to a maximum near the middle of the channel). Consider now the Péclet number. A typical molecular diffusion coefficient ranges between 10^{-5} cm^2/s at the high end (corresponding to a small molecule) and 10^{-7} cm^2/s at the low end (typical of large molecules; e.g. haemoglobin in water corresponds to 10^{-7} cm^2/s). Thus, the typical values of advective to diffusional time scales range between 10^3 and 10^5 indicating that advection is much faster than molecular diffusion. Thus, in spite of the small dimensions, molecular diffusion may not be counted on to homogenize the system to molecular scales in a reasonable amount of time. This can be seen also by calculating time required for diffusion, t_D (i.e., neglecting advection) to move a particle the width of the channel, $t_D \sim h^2/D$. This is 40 seconds for $D \sim 10^{-5}\,\mathrm{cm}^2/\mathrm{s}$, to about one hour for $D \sim 10^{-7}\,\mathrm{cm}^2/\mathrm{s}$.

Example: more on channel flow

Suppose that the two entering fluid streams flowing side by side in the channel are miscible. Then molecular diffusion provides a mechanism for the streams to penetrate into each other. The distance of penetration of one stream into another due to diffusion, δ_D, at time t, is $\delta_D \sim (Dt)^{1/2}$. Both fluids occupy the entire width of the channel after they have flowed a distance Ut_D down the channel. This distance ranges from 40 cm to 4000 cm depending on the value of D. These distances may be prohibitively long for typical microfluidic applications.

These estimates lead to three related observations important in channel flows:

- First, let us revisit the notion of 'penetration distance' discussed above from an alternate point of view. As we have seen, to reach $\delta_x = h$ solely relying on molecular diffusion takes a time $\sim h^2/D$. So if the streams move with speed U this process will have occurred after the streams have flowed a distance $L \sim U(h^2/D)$ along the channel (i.e., in the z direction). From the definition of Péclet number given above, this gives $L/h \sim \mathrm{Pe}$. Given the typical (large) values of Pe, this may be unacceptably high for microfluidic applications.
- The second observation is that as diffusion takes place in the cross-section of the channel (the plane x–y), particles experience a range of velocities (recall that the flow is parabolic), resulting in concentration dispersion in the z-direction and in a dispersion coefficient (Taylor dispersion) that scales as $1/D$. This means that fluid that disperses slowly in the cross-section will disperse rapidly in the z-direction, and vice versa.
- The third and final observation is also a consequence of the parabolic nature of the velocity field. The residence time distribution is a standard diagnostic for quantifying mixing in channel flows. Roughly, it is a probability density function consisting of the number of particles that reach the end of the channel in a given time. Near the wall the velocity field is linear with distance, $v_z \sim \dot{\gamma}d$, and thus a particle a distance d away from the wall takes a time $L/(\dot{\gamma}d)$ to reach L (hence $\dot{\gamma}$ is the shear rate at the wall). Therefore particles near the wall (as $d \to 0$) take a long time to reach the end of the channel. This would result in 'long tails' in the residence time

distribution (RTD) for particles in the channel. Moreover the fluid near the wall never co-mingles with fluid elements in the centre of the channel. The result is that mixing is poor.

Putting all this together, it is then clear that the key to effective mixing in a channel lies in the ability to mix material in the *cross-section* – to create a large amount of contact interface between the two fluids. Material 'sticking' to walls is bad for mixing. Two advantages come with enhanced mixing in the cross-section. The first is that if particles explore all of the cross-section (i.e., x–y space) in a random manner they will experience all velocities (slow near the walls, fast near the centreline) and on the whole the broadening of the RTD is reduced. The second advantage has to do with transfer processes between the surface of the device and the bulk of the fluid. If mixing is effective diffusional processes are greatly accelerated; material that is near the wall goes into the bulk and vice versa, thereby eliminating a slowdown due to diminishing concentration gradients.

1.2 Stretching and folding, chaotic mixing

In the previous section there was essentially no explicit discussion of geometric aspects of the mixing of two fluids. Geometrical considerations are motivated by the fact that the objective of mixing is to produce the maximum amount of interfacial area between two initially segregated fluids in the minimum amount of time or using the least amount of energy. Creation of interfacial area is connected to stretching of lines in 2D and surface in 3D. A fluid element of length $\delta(0)$ at time zero has length $\delta(t)$ at time t; the *length stretch* is defined as $\lambda = \delta(t)/\delta(0)$; if mixing is effective λ increases nearly everywhere, though there can be regions of compression where $\lambda < 1$. In simple shear flow the fastest rate of stretching, $d\lambda/dt$, corresponds to when the element passes though the $45°$ orientation corresponding to the maximum direction of stretching in shear flow; for long times the stretching is linear in time, $\lambda \sim t$, as the element becomes aligned with the streamlines. In an elongational flow (e.g., a flow where the velocity field depends linearly on the spatial variables and contains a saddle type stagnation point) the rate of stretching is exponential, $\lambda \sim e^t$. The distance between striations is inversely proportional to the surface area and the thinner the striations the faster the diffusion. Note that the effects of stretching on accelerating diffusion enter in two different ways: more interfacial area means more area for transfer; at the same time diminishing striation thicknesses increases the concentration gradients and increases the mass flux.

In order to conceptualize the growth of interfacial area (or perimeter in the case of two dimensions), we can imagine small elements, area or line. If mixing is effective, the small elements grow in area or length (ideally, this happens everywhere in the flow; in practice some elements may get compressed). As we shall see, the striation thickness, and stretching, are related in a deep way to dynamical systems concepts – entropy, finite size Lyapunov exponents, Smale horseshoe maps (discussed in Chapter 4), and the Baker's transformation (discussed in Chapter 3).

The key to effective mixing lies in producing stretching and folding; stretching and folding may be roughly equated with chaos as we will see in later chapters. The simplest case corresponds to two dimensions. If the velocity field is steady, the mixing is poor, stretching for long times is linear, as in the case of a simple shear flow; i.e., the stretching *rate* of line elements or decays as $1/t$ (we are restricting ourselves to bounded flows; that is, we are excluding unbounded elongational flows). It is, however, relatively straightforward to produce flow fields that can generate stretching and folding and hence chaos.

Experience over the past twenty years shows that a sufficient (heuristic) condition for chaos is the 'crossing' of streamlines. That is, two successive streamline portraits, say at t and $t + \Delta t$ for time periodic two-dimensional flows, or at z and $z + \Delta z$ for spatially periodic flows, when superimposed, should show intersecting streamlines when projected onto the x–y plane. In two-dimensional systems this can be achieved by time modulation of the flow field, for example by motions of boundaries or time periodic changes in geometry. In this monograph we show that this criterion is encapsulated by *linked twist maps* (LTMs). Figure 1.2 from Ottino & Wiggins (2004) shows a schematic representation of a channel type micromixer constructed from the concatenation of basic mixing elements. In this illustration we consider the minimal number of different mixing elements, two. Cross-sectional streamline patterns at the end of each mixing element are shown. The details of the shape and internal structure of the channel are purposefully not shown. The point here is that they can be anything that produces the desired cross-sectional flow. We illustrate the mixing properties by placing red and blue 'blobs' at the beginning of the mixer and observing how they mix as they travel down the length of the mixer. This mixer can be analyzed with the LTM formalism, which provides sufficient conditions for (mathematically) optimal mixing. It is significant to note that a (seemingly) slight change in the streamline patterns can lead to a dramatic change in the mixing properties.

Numerous experimental studies have revealed the structure of chaotic flows. The most studied cases correspond to time-periodic flows. Dye structures of passive tracers placed in time-periodic chaotic flows evolve in an iterative

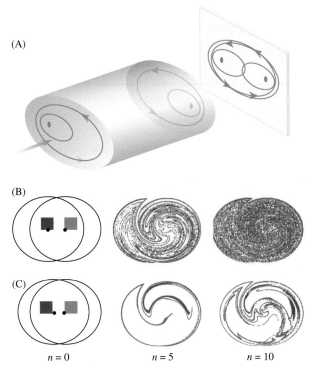

Figure 1.2 (A) Schematic representation of a channel type micromixer constructed from the concatenation of basic mixing elements. (B) The LTM mechanism causes the flow to mix completely after passing through five periodic elements of the mixer (where each consists of two of the basic mixing elements). (C) The LTM conditions are not satisfied and the flow exhibits islands, which result in poor, and incomplete mixing. [Figure taken from Ottino & Wiggins (2004).]

fashion; an entire structure is mapped into a new structure with persistent large scale features, but finer and finer scale features are revealed at each period of the flow. After a few periods, strategically placed blobs of passive tracer reveal patterns that serve as templates for subsequent stretching and folding. Repeated action by the flow generates a lamellar structure consisting of stretched and folded striations, with thicknesses $s(t)$, characterized by a probability density function, $f(s,t)$, whose mean, on the average, decreases with time. The striated pattern quickly develops into a time-evolving complex morphology of poorly mixed regions of fluid (islands) and of well-mixed or chaotic regions. Islands translate, stretch, and contract periodically and undergo a net rotation, preserving their identity returning to their original locations. Stretching within islands, on average, grows linearly and much slower than in chaotic regions, in

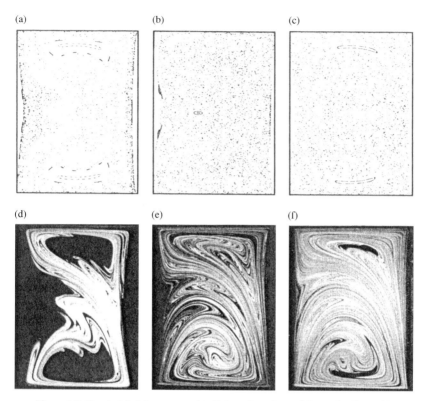

Figure 1.3 Panels (a)–(c) correspond to Poincaré sections of the cavity flow with three different protocols for the motion of the top and bottom boundaries. Immediately below each Poincaré section is a dye advection pattern for the same protocol. [Figure taken from Jana *et al.* (1994b).]

which the stretching increases exponentially with time. Moreover, since islands do not exchange matter with the rest of the fluid (in the absence of diffusion) they represent an obstacle to efficient mixing. Figure 1.3 from Jana *et al.* (1994b) shows Poincaré sections and dye advection patterns in a cavity. The flow is driven by moving the top and bottom boundaries according to a defined *protocol*. Three different protocols are shown, and each results in a different mixing pattern. By comparing the Poincaré sections to the dye advection patterns one easily sees that islands lead to poor mixing and chaos corresponds to 'good' mixing.

Now we consider a few aspects of mixing in a channel-like device: a duct flow. Duct flows are a basic configuration for many mixing devices. However, like steady two-dimensional flows, they are poor mixers. More precisely, duct flows are defined by the following velocity field

$$v_x = \frac{\partial \psi}{\partial y}, \qquad v_y = -\frac{\partial \psi}{\partial x}, \qquad v_z = f(x,y).$$

That is, a duct flow is a two-dimensional cross-sectional flow augmented by a unidirectional axial flow. Note that in a duct flow, the cross-sectional and axial flows are independent of both time and distance along the duct axis.

Duct flows can be converted into efficient mixing flows (i.e., flows with an exponential stretch of material lines with time) by time-modulation or by spatial changes along the duct axis. One example of the spatially periodic class, is the classical partitioned pipe mixer (PPM). This flow consists of a pipe partitioned with a sequence of n orthogonally placed rectangular plates. The cross-sectional motion is induced through rotation of the pipe with respect to the assembly of plates whereas the axial flow is caused by a pressure gradient. There is one control parameter in the system: ratio of cross-sectional twist to mean axial flow, β (Khakhar *et al.* (1987), Kusch & Ottino (1992)). The flow is regular for no cross-sectional twist ($\beta = 0$), and becomes chaotic with increasing values of β. In Figure 1.4 we show Poincaré sections from Khakhar *et al.* (1987) for different values of β. The Poincaré sections are obtained by mapping particles under the flow from the cross-section of the flow at the beginning of one mixing element to the beginning of the next (see also Section 2.6). Notice how dramatically the distribution and sizes of islands and chaotic regions can change with β.

To give a few typical numbers, consider a striation thickness reduction, or equivalently length stretch, where the initial length scales $s(0) \sim h$ is reduced to a size $s(t_F)$ in an amount of time t_F. According to the typical numbers given earlier we take the typical shear rates in our device to be $\dot{\gamma} = U/(h/2) \sim 10^2 \, \mathrm{s}^{-1}$. Consider a typical striation thickness reduction $s(0)/s(t_F)$ or length stretch $\lambda \sim 10^4$; that is a reduction from $10^2 \, \mu\mathrm{m}$ to $10^{-2} \, \mu\mathrm{m}$ or $10\,\mathrm{nm}$. At $10\,\mathrm{nm}$ molecular diffusion is fast at these scales, $10^{-7} \, \mathrm{s}$ for $D = 10^{-5} \, \mathrm{cm}^2/\mathrm{s}$, to $10^{-5} \, \mathrm{s}$ for $D = 10^{-7} \, \mathrm{cm}^2/\mathrm{s}$.

How long does it take to accomplish this striation thickness reduction? In simple shear, we have that $s(0)/s(t_F) \sim \dot{\gamma} t_F$; therefore the time needed to accomplish this reduction is $10^4/10^2 \, \mathrm{s}^{-1} = 10^2 \, \mathrm{s}$. An elongational flow on the other hand can accomplish the same reduction with a much lower value of elongational rate as compared with $\dot{\gamma}$; in this case $s(0)/s(t_F) = e^{\alpha t_F}$. Thus $\alpha = \ln(10^4)/100\,\mathrm{s} \sim 4 \times 10^{-2} \, \mathrm{s}^{-1}$. Elongational flows are not practical; however a succession of simple shear flows with a periodic reorientation of the line elements accomplishes the same objective.

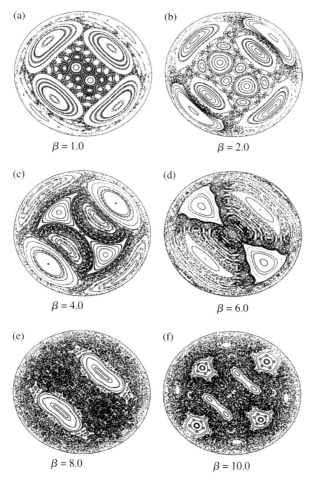

(a) $\beta = 1.0$

(b) $\beta = 2.0$

(c) $\beta = 4.0$

(d) $\beta = 6.0$

(e) $\beta = 8.0$

(f) $\beta = 10.0$

Figure 1.4 Poincaré sections for the partitioned-pipe mixer for different values of β. [Figure taken from Khakhar *et al.* (1987).]

1.3 Reorientation

Many chaotic flows may be imagined as a sequence of shear-like flows with time-periodic random reorientations of material elements relative to the flow streamlines. In all cases, the effect of the reorientation is an exponential stretching of material elements; roughly, the total stretch is the product of the stretching in each element (see Equation (1.1) below). The interval between two successive reorientations is an important parameter of such systems. In general there is an optimum interval such that the total length stretch is maximum for a fixed time of mixing. In the limit of very small periods, material elements are stretched and compressed at random, and hence the average length stretch is small (and there is an unnecessarily large amount of energy expenditure). In the limit of very large time periods, the flow approaches a steady shear flow and again the total length stretch is small. The maximum in the average stretching efficiency for simple shear flows and vortical flows corresponds when the strain per period is between 4 and 5 (Ottino (1989a)). Similar results are obtained when there is a distribution of shear rates.

The discussion so far has been in terms of average striation thickness; in practice there is a distribution of values. Computational studies indicate that within chaotic regions, the distribution of stretches becomes self-similar, achieving a scaling limit and approaches a log-normal distribution at large n. A rough argument is as follows. Let $\lambda_{n,n+1}$ denote the length stretch experienced by a fluid element between periods n and $n+1$. The total stretching after m periods of the flow, $\lambda_{0,m}$, can be written as the product of the stretchings from each individual period:

$$\lambda_{0,m} = \lambda_{0,1}\lambda_{1,2} \cdots \lambda_{m-1,m}. \tag{1.1}$$

The stretchings between successive periods (i.e., $\lambda_{1,2}$ and $\lambda_{2,3}$) are strongly correlated, however. The correlation in stretching between non-consecutive periods (*e.g.*, $\lambda_{0,1}$ and $\lambda_{4,5}$) grows weaker as the separation between periods increases due to chaos (the presence of islands in the flow complicates the picture). Thus, $\lambda_{0,m}$ is essentially the product of random numbers and

$$\log \lambda_{0,m} = \log \lambda_{0,1} + \log \lambda_{1,2} + \cdots + \log \lambda_{m-1,m}.$$

is a sum of random numbers. According to the central limit theorem, any collection of sums of random numbers will converge to a Gaussian. Therefore the distribution of $\lambda_{0,m}$ is log-normal.

1.4 Diffusion and scaling

In the context of the discussion above, we now consider the role of molecular diffusion. Consider molecular diffusion across a thinning striation with striation thickness $s(t)$ as it is followed in a Lagrangian sense along a mixer (the arguments are similar to those in Ottino (1994); we correct a couple of typographical errors in the original paper). The initial thickness, $s(0) \sim h$, is thinned down according to a stretching function $\alpha(t)$ given by $d\ln(s(t))/dt = -\alpha(t)$. The stretching function is bounded by the shear rate; in chaotic flow the time average of α is positive; in two-dimensional flows or duct flows it decays as

$1/t$. As a rule of thumb the value of α is typically an order of magnitude smaller than the typical shear rate, U/h.

In the frame of the striations the diffusion process is described by

$$\frac{\partial c}{\partial \tau} = \frac{\partial^2 c}{\partial \xi^2},$$

where c is the concentration, ξ is a striation-thickness-based spatial coordinate normal to the striations, and τ is the so-called warped time, defined as:

$$\xi = \frac{x}{s(t)},$$

and

$$\tau = \int \frac{D}{(s(t'))^2} dt'.$$

The penetration distance in the (ξ, τ) space is given by $\delta_\xi \sim \tau^{1/2}$ and therefore in terms of x, t variables we have

$$\frac{\delta_x}{s(0)e^{-\alpha t}} = \left[\frac{D}{(s(0))^2 2\alpha} (e^{2\alpha t} - 1) \right]^{1/2}.$$

Thus, for long times the penetration distance stabilizes to $\delta_x \sim (D/\alpha)^{1/2}$. This time may not be reached in practice, however, as striations fuse together due to molecular diffusion. One can argue that the mixing is complete when $\delta_x = s_f$, the penetration distance growth catches up with the thinning striations after a time t_F. This happens when

$$1 = \left[\frac{D}{(s(0))^2 2\alpha} (e^{2\alpha t} - 1) \right]^{1/2}.$$

The value of α can be estimated as the inverse of the shear rate, i.e. $\alpha \sim U/h$ therefore $\mathrm{Pe} \sim \alpha h^2 / D$. Therefore, if $\exp(2\alpha t_F) \gg 1$, the necessary length for mixing scales as

$$\frac{L}{h} \sim \log \mathrm{Pe}.$$

Clearly this is much more efficient than the $L/h \sim \mathrm{Pe}$ relationship uncovered earlier.

1.5 Examples

In this section we will describe a collection of examples that come from diverse areas of applications that span a wide range of length and time scales. Nevertheless, all of the examples embody the paradigm of 'crossing of streamlines.'

In the next chapter we will see that this paradigm can be realized in a rigorous, mathematical framework as a *linked twist map*.

1.5.1 The Aref blinking vortex flow

The 'blinking vortex flow' was introduced by Aref (1984), with further work by Khakhar *et al.* (1986). This is a seminal example in the field of chaotic advection. It is the flow generated by a pair of point vortices separated by a finite distance, that blink on and off periodically in an unbounded inviscid fluid. At any given time only one of the vortices is on so that the motion of a fluid particle during a period is made up of two consecutive rotations about different centres.

The velocity field due to a single point vortex located at the position $(a, 0)$ in a Cartesian coordinate system is given by (in polar coordinates)

$$\dot{r} = 0,$$

$$\dot{\theta} = \frac{\Gamma}{2\pi r},$$

where Γ is the strength of the vortex and $r = \sqrt{(x-a)^2 + y^2}$. The velocity field can easily be integrated over the time t for which the vortex exists to obtain the following *twist map*:

$$T(r, \theta) = (r, \theta + 2\pi g(r)),$$

where

$$g(r) = \frac{\Gamma t}{4\pi^2 r}.$$

Now consider two identical point vortices located at $(-a, 0)$ and $(a, 0)$. We imagine the situation where the vortex at $(-a, 0)$ exists for time t, turns off, then the vortex at $(a, 0)$ turns on and exists for a time t, turns off, with the process repeating in this way. We denote the twist map at $(a, 0)$ by T_+ and the twist map at $(-a, 0)$ by T_- (which is obtained from T_+ by letting $a \to -a$). We will see in the next chapter that the evolution of a fluid particle is governed by the *linked twist map* $T = T_+ \circ T_-$.

1.5.2 Samelson's tidal vortex advection model

The 'blinking vortex flow' has been used to model a variety of flows. Here we describe a type of blinking flow example that arises in a geophysical fluid dynamics setting. Consider the situation of an ingoing and outgoing flow due to the tides along a segment of shoreline having a headland. It has been observed

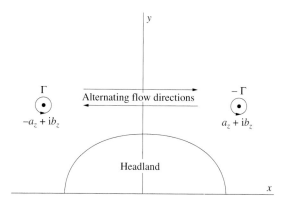

Figure 1.5 Illustration of Samelson's tidal vortex advection model.

(Signell & Geyer (1991)) that during this process eddies are sequentially generated on opposite sides of the headland. Understanding the mixing processes in such situations is important for understanding a variety of environmental and biological processes occurring in such coastal settings (see e.g., Signell & Butman (1992)). Samelson (1994) has developed a kinematic model to study mixing and transport by eddies shed from a headland as a result of tidal flow, which we now briefly describe (see Figure 1.5).

The domain for Samelson's model consists of a straight boundary along $y = 0$ with a semicircular headland of radius 1 centered at the origin. The flow is generated by a sequence of four flows:

1. Tidal flow modelled by a constant translation in the positive x direction.
2. Eddy advection modelled by a point vortex of strength $-\Gamma$ located at $a_z + ib_z$.
3. Reverse tidal flow modelled by a constant translation in the negative x direction.
4. Eddy advection modelled by a point vortex of strength $+\Gamma$ located at $-a_z + ib_z$.

Samelson shows, through a sequence of conformal maps, that this problem can be mapped directly onto the blinking vortex flow considered by Aref (1984).

1.5.3 Chaotic stirring in tidal systems

Here we describe another example that we feel can be studied fruitfully via the linked twist map framework that we develop in this book (although some crucial modifications will be required). Ridderinkhof & Zimmerman (1992),

building on earlier work of Pasmanter (1988), developed a model to describe
mixing in a tidal basin that exhibits 'streamline crossing' as we have described.

Their model consists of the superposition of a tidal field, assumed to be a
spatially uniform oscillating current in one direction, and a residual current
consisting of an infinite sequence of clockwise and counterclockwise eddies.
The model is realized with the following (dimensionless) streamfunction:

$$\psi(x, y, t) = \lambda T(t) y + \frac{\lambda v}{\pi} [1 + T(t)] \sin \pi x \sin \pi y,$$

from which the following velocity field is obtained:

$$\dot{x} = \frac{\partial \psi}{\partial y} = \lambda T(t) + \lambda v [1 + T(t)] \sin \pi x \cos \pi y,$$

$$\dot{y} = -\frac{\partial \psi}{\partial x} = -\lambda v [1 + T(t)] \cos \pi x \sin \pi y.$$

The dimensionless parameter λ is the ratio of the tidal excursion to the eddy
diameter and the dimensionless parameter v is the ratio of the residual eddy
velocity to the tidal velocity amplitude. Both parameters are positive with λ
ranging up to 4 and v up to 0.32 (see Beerens *et al.* (1994) for a thorough
discussion of the physical origin of these parameters and their values). The time-
dependence is given by the following function (Ridderinkhof & Zimmerman
(1992)):

$$T(t) = \begin{cases} 1 & \text{for} \quad k < t \leq k + \tfrac{1}{2}, \\ -1 & \text{for} \quad k + \tfrac{1}{2} < t \leq k + 1. \end{cases}$$

Hence, the model is a 'blinking flow'. More explicitly, during each half cycle
the particles are advected by the following two velocity fields:

$$\dot{x} = \lambda + 2\lambda v \sin \pi x \cos \pi y,$$

$$\dot{y} = -2\lambda v \cos \pi x \sin \pi y, \qquad k < t \leq k + \frac{1}{2}, \qquad (1.2)$$

and

$$\dot{x} = -\lambda,$$

$$\dot{y} = 0, \qquad k + \frac{1}{2} < t \leq k + 1 \qquad (1.3)$$

In Figure 1.6 we show the streamlines of these two velocity fields for one spatial
period. If one superimposes Figure 1.6(a) and Figure 1.6(b) one clearly sees
the phenomenon of streamline crossing.

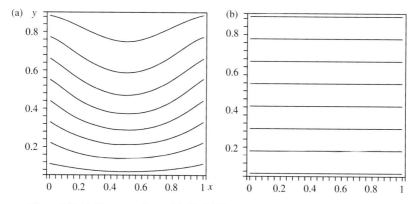

Figure 1.6 (a) The streamlines of (1.2). (b) The streamlines of (1.3). We have taken $\lambda = 0.25$, $\nu = 0.31$.

If we let M_1 denote the map obtained by letting particles flow for a time $1/2$ under the velocity field (1.2) and M_2 denote the map obtained by letting particles flow for a time $1/2$ under the velocity field (1.3), then the advection of fluid particles is described by iteration of the map $M = M_2 \circ M_1$. As a consequence of the spatial periodicity in both directions this map is defined on the two-dimensional torus. While M has the structure of the composition of two 'linked' maps on the torus, it does not fit into the standard linked twist map framework that we will describe shortly. Nevertheless, we believe it is a new generalization that may be studied fruitfully by the same approach, and yield some interesting new dynamics at the same time. One issue is that the shear exhibited in Equation (1.3) is constant, i.e. it is the same on each streamline. For classical twist maps the particle speed will vary from streamline to streamline.

1.5.4 Cavity flows

The flows produced in a region bounded by two opposing non-moving and two opposing moving walls are referred to as cavity flows. This class of flows have become an archetypal flow for experimental and computational studies of chaotic flows. They were introduced by Chien *et al.* (1986), and developed further by Leong & Ottino (1989) and Jana *et al.* (1994b). The mode of operation is typically at low Reynolds numbers, so the streamline portraits contain no information as to the direction of the flow (that is, all directions of the flow can be reversed and the streamlines would not change). When operated in a blinking mode (one wall is moved for a certain distance and then stopped, and

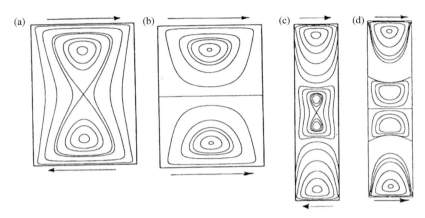

Figure 1.7 Streamline patterns in a cavity for different aspect ratios and boundary motions (the arrows indicate different directions of the motion of the boundary). [Figure taken from Jana *et al.* (1994b).]

then another wall is moved for a certain distance and then stopped), the system can be interpreted as an LTM. Depending on the aspect ratio, shape of the cavity (walls need not be parallel), and the addition of internal baffles, one can generate a wide variety of streamline patterns, some of which are shown in Figure 1.7, from Jana *et al.* (1994b).

1.5.5 An electro-osmotic driven micromixer blinking flow

Qian & Bau (2002) considered flows generated by electro-osmotic flow (EOF) in cavities. Until recently EOFs have been used primarily as an alternative to pressure-driven flow in microchannels, the simplest case corresponding to uniformly charged walls. However, several other scenarios are possible; an early study considering the effects on nonuniform charge is by Anderson & Idol (1985). Qian & Bau (2002) computed flow patterns for specific (nonuniform) potential distributions on the walls of the cavity. Different potential distributions gave rise to different cellular flow fields in the cavity, as shown in Figure 1.8.

Qian and Bau also suggested that one could switch between different flow patterns through 'judicious control of embedded electrodes' in the walls of the cavity. In this way a blinking flow can be realized. Not surprisingly, they demonstrated numerically that such flows can give rise to chaotic fluid particle trajectories. Clearly, such a scheme also fits squarely within the LTM formalism. If we superimpose two chosen flow patterns that are rigid rotations of

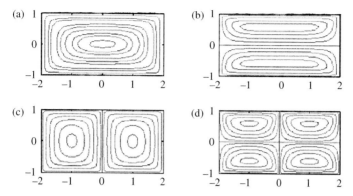

Figure 1.8 Figure 4 from Qian & Bau (2002) showing flow patterns they computed for different ζ potential distributions on the walls of the cavity.

each other the structure of the linked twist map is clear. One way of applying the results on LTMs is to choose annuli in one flow pattern and other annuli in the other flow pattern such that the annuli intersect pairwise 'transversely' in two disjoint components. Then the switching time between patterns, T, is chosen such that for each annulus the outer circle rotates twice with respect to the inner circle during the time. We then need the twists to be 'sufficiently strong', which will also depend on whether or not the chosen annuli pair are co- or counter-rotating. If this can be done, then appealing to the dynamical systems results described later, the flow will have 'strong mixing properties' in the region defined by the chosen annuli. Of course, there are numerous open problems. For example, which potential distributions lead to the maximal region on which the linked twist map results hold? However, such an analysis is possible using formulae for the flow patterns given by Qian & Bau (2002).

1.5.6 Egg beater flows

Franjione & Ottino (1992) developed a model that encapsulates the essential kinematic mechanisms for 'good mixing' in a large class of mixers. The model was arrived at as a result of the accumulation of a great deal of experience analyzing a variety of diverse mixing situations from the dynamical systems point of view. Interestingly, and unrecognized at the time, the model is precisely a linked twist map on the torus.

(a)

(b)

(c)

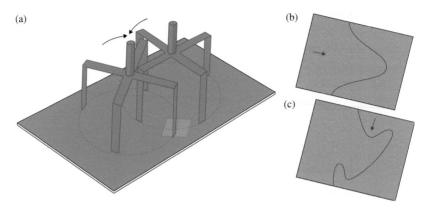

Figure 1.9 (a) Schematic of the egg beater. In (b) a blade pushes a material line, in (c) a second blade folds the line. [Figure from Ottino (1989b).]

We consider a unidirectional flow of the following form:

$$\frac{dx}{dt} = v(y),$$

$$\frac{dy}{dt} = 0.$$

These equations are easily integrated to obtain the following trajectories:

$$x = X + v(Y)t,$$

$$y = Y,$$

where (X, Y) denote the position of the fluid particle at $t = 0$. At this point we have not specified a velocity profile $v(y)$. However, regardless of the form of $v(y)$, the mixing will be poor since the flow is steady, trajectories cannot cross, and therefore material will tend to align with the x-axis. If we alter this situation after a certain time by rotating the system by 90° then material that was previously oriented parallel to the streamlines is now normal to the (rotated) streamlines. This procedure of rotating the flow 'back and forth' between the original flow and one oriented at a 90° angle with respect to the original can be repeated indefinitely. Franjione & Ottino (1992) refer to this sequence of orthogonally oriented flows as the *egg beater flow* since it represents a simplified picture of the mixing mechanism in a hand held egg beater (see Figure 1.9). In an egg beater there is no loss of material, and this is incorporated in this model by assuming that the flow occurs in a domain that is periodic in both the x and y directions. Hence, we consider the domain of the flow to be given by the box $0 \leq x \leq 1$ and $0 \leq y \leq 1$ and when a particle exits one side of the box it re-enters at the same point on the opposite side. In terms of the physical model of the egg beater, this can be interpreted to be that whenever a blade of the egg beater leaves one side of the domain, a new blade enters on the opposite side. Mathematically, the flow is said to take place on a torus.

The flow described above can be more precisely expressed as a mapping. Using the expression for the trajectories given above, a particle at (x_n, y_n) is mapped in the horizontal direction after a time T by:

$$x_{n+1} = x_n + Tv(y_n),$$
$$y_{n+1} = y_n, \qquad\qquad (1.4)$$

where we can think of T as the length of time that the blade moves in the x direction. Adopting a shorthand vector notation, we denote $\mathbf{x} = (x, y)$ and the map by $\mathbf{H}$ so that (1.4) becomes:

$$\mathbf{x}_{n+1} = \mathbf{H}\mathbf{x}_n.$$

The second mapping in the vertical direction is similarly easily found to be:

$$x_{n+1} = x_n,$$
$$y_{n+1} = y_n + Tv(x_n),$$

and we adopt a similar vector notation to write the map more succinctly as:

$$\mathbf{x}_{n+1} = \mathbf{V}\mathbf{x}_n.$$

The overall mapping describing the flow is then given by the composition of these two maps:

$$\mathbf{x}_{n+1} = \mathbf{V}\mathbf{H}\mathbf{x}_n.$$

Of course, there is considerable scope for generalization here. For example, the horizontal and vertical flows could have different velocity profiles, and they could also act for different times.

Figure 1.9 from Ottino (1989b) illustrates the key kinematical features of the egg beater. Note the highlighted square in panel (a). A fluid line element in the square perpendicular to a blade is deformed as shown in panel (b) as a blade pushes through it. The other blade pushes through the line in a perpendicular direction, as shown in panel (c). Parts of the line element that extend out the top of the square later re-enter through the bottom.

If the blades are rotated alternately then the flow can be described by a LTM. However, this is a LTM *on the plane* rather than a torus. Also, note that the two blades are rotating in the opposite sense, i.e., the one on the left is rotating clockwise and the one on the right is rotating counterclockwise. Fluid mechanicians might refer to this as a counter-rotating situation. However, note that the blades are pushing through the highlighted square of fluid in the same sense. This point is discussed further in Section 2.3.1.

1.5.7 A blinking flow model of mixing of granular materials

Over the past fifteen years there has been intense activity in the study of granular *flow*. However, the study of granular *mixing* has received much less attention. Mixing of granular materials is important in a variety of industrial processes (e.g., in the pharmaceutical, food, ceramic, metallurgical and construction industries), as well as natural processes such as debris flows and the formation of sedimentary structures. Specific references can be found in Ottino & Khakhar (2000). Unlike fluids, the flow in a mixer does not always lead to mixing of the material. For example, particles can segregate as a consequence of different particle properties (e.g. size differences and density differences, see Jain *et al.* (2005)). Thus granular mixing provides a rich test bed for the study of pattern formation and self-organization. The lack of any fundamental theory indicates that there is tremendous scope for the development and analysis of prototypical models.

It is shown in Khakhar *et al.* (1999), Hill *et al.* (1999a), and Hill *et al.* (1999b) that the phenomena of 'streamline crossing' that we have described can occur in a large class of convex mixers (Khakhar *et al.* (1999)). Here we describe how this property can be exploited to derive a linked twist map that will describe the flow of particles in a half-full rotating tumbler.[1]

For simplicity, we begin our discussion by describing a circular tumbler, whose geometry is shown in Figure 1.10. A model for this flow under certain operating conditions is derived in Khakhar *et al.* (1997), and we briefly describe the essential points. The free surface, i.e., the boundary between the granular material and the air, is essentially a straight line, and remains at a fixed angle with respect to the horizontal. In the rolling regime[2] the flow domain is divided

---

[1] The shape of the mixer that we describe, e.g., circle or square, is that of the cross-section of a long channel, where every cross-section has the same shape. Hence, we will be discussing mixing in the cross-section or 'transverse mixing'. Mixing in the axial direction could also be considered. However, we will assume that there is no flow in the axial direction, and therefore the only mechanism for axial mixing would be some sort of diffusive process, but that is an effect that we will not consider here.

[2] It is not hard to imagine that if the rotation rate is slow, then the material will accumulate in a wedge on the left-hand side of the tumbler until it reaches a critical height, at which point an 'avalanche' occurs. For faster rotation rates this does not occur, and the material 'rolls' with the rotation of the tumbler. Still, the rotation rate is typically slower than the dynamics of the particles.

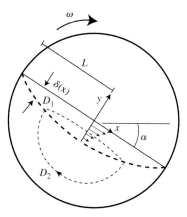

Figure 1.10 Schematic view of the continuous flow regime in a rotating cylinder. The solid straight line through the middle of the cylinder is the free surface, and this layer length is $2L$ ($\delta(0)$ is denoted δ_0). The heavy dashed curve connecting both sides of the cylinder denotes the interface between the continuously flowing layer and the region of solid body rotation. The mixer is rotated with angular velocity ω, and the velocity profile within the layer, v_x, is nearly a simple shear, as indicated in the figure. The v_y component is not shown. A typical particle trajectory is shown as a dashed closed curve.

into two distinct regions. A flowing layer is defined by:

$$D_1 = \{(x, y) \mid -L \le x \le L,\ 0 \le y < -\delta(x)\},$$

where the thickness of the flowing layer is given by:

$$\delta(x) = \delta_0 \left(1 - \left(\frac{x}{L}\right)^2 \right),$$

and a region outside this flowing layer is defined by:

$$D_2 = \{(r, \theta) \mid 0 \le r \le L,\ 0 \le y < \pi\} - D_1,$$

where the particles are assumed to undergo solid body rotation. The boundary between D_1 and D_2, denoted $\partial_{1,2}$ is therefore given by:

$$\partial_{1,2} = \{(x, y) \mid -L \le x \le L,\ y = -\delta(x)\}.$$

The phase space (which is the physical space in this case) of this dynamical system is defined on $D_1 \cup D_2$. We need therefore to define the dynamics on each region separately, and a matching condition at the boundary.

On D_1 the dynamics is given by:

$$D_1 : \begin{cases} \dot{x} = v_x = 2u\left(1 + \dfrac{y}{\delta}\right) \\[3mm] \dot{y} = v_y = -\omega x \left(\dfrac{y}{\delta}\right)^2 \end{cases}$$

where

$$u = \frac{\omega L^2}{2\delta_0}.$$

On D_2 the dynamics is given by solid body rotation.

A particle starting in D_1 or D_2 is advected by the appropriate flow. When it reaches $\partial_{1,2}$, we then have to switch the advection rule. We need a quantity to monitor along a trajectory to determine when to do this. The particle reaches $\partial_{1,2}$ when $y = -\delta(x)$. Then we need to determine which region it will enter. It suffices to monitor the x component for this simple flow. If a particle is on $\partial_{1,2}$ and $x > 0$, then it is leaving D_1 and entering D_2. If it is on $\partial_{1,2}$ and $x < 0$, then it is leaving D_2 and entering D_1. Clearly, this formulation gives an integrable model with closed 'streamlines' shown in Figure 1.11(a).

Fiedor & Ottino (2005) consider one way to operate the tumbler which breaks integrability and leads to streamline crossing, and chaotic mixing; the case of a tumbler with a circular cross-section, but where the rotation rate is varied periodically in time. As we argued above, if the cross-section of the tumbler is circular and the rotation rate is constant then the flow is steady (and integrable). However, if the rotation rate varies periodically in time this leads to a changing thickness of the flowing layer which results in a change in the streamline pattern within the flowing layer. In Figure 1.11 (b) we show streamlines at two different times (solid and dashed) that cross in the flowing layer. The different thicknesses of the flowing layer at the two different times are shown with a light and a dark shading.

The crossing of the streamlines at two different times provides the necessary structure for creating a linked twist map. The LTM context has the advantage that one could design for optimal mixing in a particular region of the flow. Here we imagine a 'blinking cylinder flow' for granular mixing that would be a model in the same way that the blinking vortex flow is for chaotic advection of fluids.

Figure 1.12 shows results from Fiedor & Ottino (2005) for the cylinder having a circular cross-section where the angular rotation rate is modulated periodically in time. The Poincaré section clearly indicates that the flow is chaotic. The experiments reveal something quite different than the experiments with fluids,

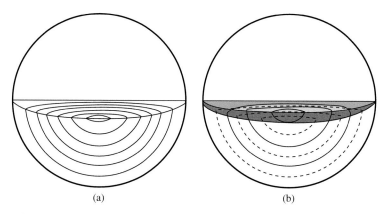

Figure 1.11 (a) Streamlines for a constant rotation rate of the cylinder. (b) Solid streamlines at one time and dashed streamlines at another time shown superimposed. Streamline crossing occurs in the flowing layer.

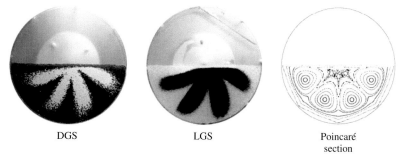

DGS   LGS   Poincaré
                                              section

Figure 1.12 Unmixing in granular materials (from Fiedor & Ottino (2005)). Time-periodic modulation in angular rotation leads to a chaotic flow structure captured in the Poincaré section of the flow. However, as opposed to fluids, continual flow leads to unmixing.

such as those of Figure 1.3. Dye experiments in chaotic mixing of fluids, primarily in time-periodic flows, have been instrumental over the last two decades in yielding insights into the working of chaotic flows. Strategically placed blobs, after a few periods of the flow, produce persistent large-scale structures – rough templates of the manifold structure – with additional periods revealing finer structures of nested striations as shown in Figure 1.3. Thus, the dyes show the chaotic regions and one 'sees' the islands as the regions where the dye does not go. Something similar may be attempted in granular flows. However, forming a blob in granular matter is hard and very quickly the blob becomes broken and connectivity of the 'dyed' structure, as opposed to the companion fluid case, is lost. Segregation experiments in granular matter, on the other hand, are easy and can be repeated multiple times. As indicated earlier a distinguishing feature of flowing granular matter is its tendency to segregate; mixtures of particles with varying size (S-systems) or varying density (D-systems) subject to flow often segregate leading to what on first viewing appear to be baffling results. This phenomenon occurs in dry granular systems (DGSs) and liquid granular systems (LGSs; i.e., DGSs where all air is replaced by a liquid). One starts with a well-mixed system and 'mixing' leads to unmixing, as shown in Figure 1.12. After the experiment is done, one remixes the system by shaking (one has to be careful not to unmix) and a new experiment is ready to go. The experiments of Figure 1.12 indicate that LGSs and DGSs produce similar segregation patterns. In the DGS case the smaller particles are white and in the LGS the small particles are black. Thus, as opposed to the case of mixing of fluids, the particles tag the location of the largest islands in the flow. This is an area of active investigation at the present time.

1.5.8 Mixing in DNA microarrays

The flows used in DNA microarrays display the signature of crossing of streamlines. The exploitation of this may be very significant since effective mixing is crucial for the functioning of these devices. DNA microarrays are now an essential tool for obtaining genetic information, and the key process in obtaining this information is DNA hybridization. DNA hybridization is a mixing process where speed is essential for a variety of reasons. DNA hybridization occurs in a large aspect ratio (i.e., 'thin') mixing chamber with horizontal dimensions of the order of 10 mm and depth of the order of 0.5 mm. The bottom of the chamber consists of an array of probes each having an oligonucleotide (i.e. a small DNA molecule composed of a few nucleotide bases) having a specified sequence. A solution of labelled DNA is introduced into the mixing chamber and when the labelled DNA combines with its complementary sequence on a probe, the probe is said to be hybridized. The hybridized DNA can then be studied to determine the degree of genetic similarity of the two species. In order to achieve a larger sample of hybridized DNA it is important that the DNA samples can interact equally with all probes in the array. Typically, molecular diffusion has been relied upon as the mechanism for achieving this interaction. However, recently it has been recognized that chaotic advection can make the process much more efficient (McQuain *et al.* (2004), Raynal *et al.* (2004)). Here we show that the mixing process in a DNA microarray can be modelled as a type of toral LTM.

In Figure 1.13 we show a schematic view of the hybridization chamber taken from Raynal *et al.* (2004). Fluid flow in the chamber is induced by creating a pressure differential between opposite holes. Practically speaking this means

Figure 1.13 Schematic of the mixing chamber for the DNA microarray. The shaded region contains the DNA probes. The black circles are points at which fluid is introduced and extracted from the mixing chamber.

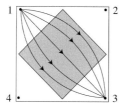

Figure 1.14 Streamlines corresponding to fluid entering the chamber from 1 and exiting through 3. The exiting fluid is re-injected into 1.

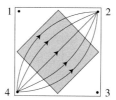

Figure 1.15 Streamlines corresponding to fluid entering the chamber from 4 and exiting through 2. The exiting fluid is re-injected into 4.

that with a system of syringes and tubes fluid is:

1. injected from 1, and ejected into 3,
2. injected from 4 and ejected into 2,

and these two steps are repeated periodically.

Hole 1 is a source of fluid and hole 3 is a sink. Fluid that goes into the sink (hole 3) is re-injected into the source (hole 1). The flow induced by a source and sink can be solved exactly, and is completely integrable. The streamlines for this 'source-sink pair' are shown in Figure 1.14. Similarly, hole 4 is a source and hole 2 is a sink. The streamlines for this 'source-sink pair' are shown in Figure 1.15.

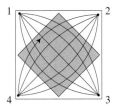

Figure 1.16 The crossing streamlines of the two different source-sink pairs. The 'crossing' occurs at two different times., i.e. the two different half-cycles of the mixer.

This flow can be modelled as a pair of alternating source-sink pairs – a type of blinking flow. (Chaotic advection in a pulsed source-sink pair was studied in Jones & Aref (1988).) The first source sink pair (i.e. 1 and 3) operates for time T/2. Fluid flowing into the sink is re-injected into the source during this time. After T/2 the other source-sink pair (i.e. 2 and 4) is 'turned on' for time T/2. The streamlines for each half period 'cross', as shown in Figure 1.16.

This situation is interesting because it gives the structure of a linked twist map on a torus. This can be seen as follows. Because fluid that exits a sink is re-injected from the source at which it entered, we can view the source-sink pair as 'identified' in just the same way that we identify the vertical edges of a square of a fixed length to obtain periodic boundary conditions. We show this in Figure 1.17.

Of course, there is a slight complication with this picture. In this analogy the vertical sides of a square are being collapsed to a point. In other words, if a fluid particle goes down a sink, what 'direction' does it come out of the source? We must adopt some rule for this (e.g. it exits on a given streamline, at a random angle, etc.). Also, the standard 'twist condition' on the 'centreline' connecting source-sink pairs breaks down. Nevertheless, the LTM framework provides a framework, and a variety of tools, for rigorous mixing studies of this system.

1.6 Mixing at the microscale

Mixing at the microscale is an increasingly important subject that can be analysed in detail by the methods developed in this book. Microfluidics is the term that is used to describe flow in devices having dimensions ranging from millimeters to micrometers and capable of handling volumes of fluid in the range of nano to micro litres (10^{-9}–10^{-6} L).

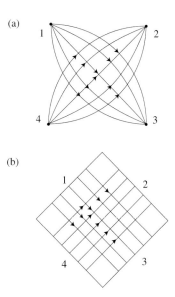

Figure 1.17 (a) The crossing streamlines of the two different source-sink pairs. (b) The 'crossing' streamlines after blowing up hole 1 and hole 3 into to 'horizontal' lines (although here we draw the square in the same alignment as the original), and then identifying them in the usual torus construction (i.e., periodic boundary conditions). Similarly, holes 2 and 4 can be blown up into 'vertical' lines.

Mixing – or lack thereof – is often crucial to the effective functioning of microfluidic devices (Knight (2002)). Often the objective is rapid mixing between two initially segregated streams – rapid interspersion in the minimal amount of space. Other times, however, the objective is to prevent mixing and maintain segregation; for example, having two streams co-flowing side by side and controlling or monitoring processes occurring at the interface between the two fluids. There are excellent reviews of general aspects of microfluidics and mixing is covered to various degrees in several of them (see for example Stone *et al.* (2004) and Stroock & Whitesides (2001)). However, there appears to be no review wholly devoted to mixing – how to enhance, how to control it, or simply how to benefit from existing theory.

The key aspect about microfluidics is smallness, and smallness brings new elements, not only quantitative, but also qualitative. The role of interfaces becomes dominant. In solid-fluid interfaces, wettability and charge density can be exploited in various ways; in the case of liquid–liquid or liquid–gas interfaces, gradients of interfacial tension can have large effects. A lot of the basic science involved in these developments is already established. What is important however is the possibility of invention of new designs by exploiting boundary conditions that are simply ineffective at larger scales. Surface patterning – a wall with small grooves oriented at oblique angles with respect to the axis of the main flow (e.g. Stroock *et al.* (2002)) – suggests several possible designs, as we show in Section 2.6.

2

Linked twist maps: definition, construction, and the relevance to mixing

In this chapter we give formal definitions of linked twist maps on the plane and linked twist maps on the torus. We give heuristic descriptions of the mechanisms that give rise to good mixing for linked twist maps, and highlight the role played by 'co-rotation' and 'counter-rotation'. We show how to construct linked twist maps from blinking flows and from duct flows, and we describe a number of additional examples of mixers that can be treated within the linked twist map framework.

2.1 Introduction

The central theme of this book is that the mathematical notion of a *linked twist map*, and attendant dynamical consequences, is naturally present in a variety of different mixing situations. In this chapter we will define what we mean by a linked twist map, and then give a general idea of why they capture the essence of 'good mixing'. To do this we will first describe the notion of a linked twist map as first studied in the mathematical literature. This setting may at first appear to have little to do with the types of situations arising in fluid mechanics, but we will argue the contrary later. However, this more mathematically ideal setting allows one to rigorously prove strong mixing properties in a rather direct fashion that would likely be impossible for the types of maps arising in typical fluid mechanical situations. We will then consider a variety of mixers and mixing situations and show how the linked twist map structure naturally arises. Most importantly, if one considers the common geometrical features of the streamlines of each of the mixers and mixing situations that we consider it will be clear that linked twist maps embody a paradigm for strong mixing properties, namely, *streamline crossing*. This was first pointed out in Ottino (1990), and it can involve streamlines crossing at different times (the blinking vortex flow, Aref (1984)) or streamline crossing at different spatial locations (the partitioned pipe mixer, Khakhar *et al.* (1987)). The abstraction of this property in the form

31

of a linked twist map is crucial because it allows, for the first time, a quantitative way to design and understand mixers that have mathematically strong mixing properties on sets of positive measure. Significantly, for the design process, attaining these strong mixing properties depends on geometrical properties of the streamlines, and *not* the manner in which those geometrical properties are realized.

2.2 Linked twist maps on the torus

Strong mixing properties of linked twist maps on a torus (sometimes referred to as 'toral LTMs') were obtained in the works of Burton & Easton (1980), Devaney (1980), Przytycki (1983), Przytycki (1986), and Wojtkowski (1980). In this section we define what is meant by a *linked twist map on the torus*. We do so fairly briefly in this section, and will give more details in Chapter 6 when we discuss precise mathematical properties.

First we consider the two-dimensional torus $\mathbb{T}^2$ with coordinates (x, y) (mod 1) (i.e., x and y are periodic of period one). On this torus we define two overlapping annuli P, Q by

$$P = \{(x, y) : 0 \leq x \leq 1, 0 \leq y_0 \leq y \leq y_1 \leq 1\}$$
$$Q = \{(x, y) : 0 \leq y \leq 1, 0 \leq x_0 \leq x \leq x_1 \leq 1\}.$$

We denote the union of the annuli by $R = P \cup Q$ and the intersection by $S = P \cap Q$, as in Figure 2.1. It is often more convenient both visually and graphically to consider the dynamics on the torus $\mathbb{T}^2$ with the torus represented as the unit square with doubly periodic boundary conditions, i.e., where the top and bottom edges of the square are identified, as are the left and right edges, as shown in Figure 2.1. The annuli P and Q then become vertical and horizontal strips in the square.

In order to define a linked twist map on the torus we first define a twist map on each annulus. A twist map is simply a map in which the orbits move along parallel lines, but with a uniform shear. In particular, we define

$$F : R \to R$$

$$F = F(x, y; f) = \begin{cases} (x + f(y), y) & \text{if } (x, y) \in P \\ (x, y) & \text{if } (x, y) \in R \backslash P \end{cases}$$

where $f : [y_0, y_1] \to \mathbb{R}$ is a real-valued function such that $f(y_0) = 0$ and $f(y_1) = k$, for some integer k, and $R \backslash P$ means 'all points in R, except for those that are also in P'. So if F acts on a point (x, y) in R that is *not* in P, it leaves

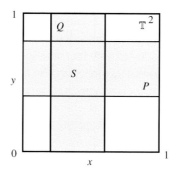

Figure 2.1 Representation of the torus on the unit square. The left and right vertical boundaries are identified and the upper and lower horizontal boundaries are identified, i.e., the domain is doubly periodic. The annuli P and Q are shown shaded.

that point unchanged (in other words, F is the *identity* map, Id, on $R\backslash P$). If F is applied to a point (x, y) in P, the y-coordinate is left unchanged, but the x-coordinate is altered by an amount dependent on the value of y. We insist that k must be an integer in order that the two components of F (i.e., the component of F defined on P and the component of F defined on $R\backslash P$) should 'join up' at the boundary of P – that is, F should be *continuous* on the boundary of P, denoted ∂P. This point is crucial mathematically and will be carefully discussed in later chapters. In terms of fluids, it corresponds to systems where initially connected blobs of fluid remain connected blobs, with no tearing or break-up taking place.

In an analogous fashion, we define a twist map on Q:

$$G : R \rightarrow R$$

$$G = G(x, y; g) = \begin{cases} (x, y + g(x)) & \text{if } (x, y) \in Q \\ (x, y) & \text{if } (x, y) \in R\backslash Q \end{cases}$$

where $g : [x_0, x_1] \rightarrow \mathbb{R}$ is a real-valued function such that $g(x_0) = 0$ and $g(x_1) = l$, for some integer l. Again $G = Id$ outside the annulus Q.

For obvious reasons we refer to k and l as the *wrapping number* of the twist. If a twist has wrapping number 1 or 2, we may also refer to it as a single-twist or double-twist respectively, again for obvious reasons. Note that in general k and l may be positive or negative integers and so twists may wrap around the torus in either direction. The choice of sign of kl makes a crucial difference to the ensuing results and methods of proof for the mixing properties of the linked twist map.

Having ensured the continuity of F and G we turn to further smoothness properties. Since the identity map is smooth (that is, infinitely times differentiable), the map F will be endowed with the same smoothness properties as the function f only if these smoothness properties also hold on ∂P. Again, this will have implications that we will discuss in detail later on.

There is much to be said about restrictions on the form of the functions f and g, and this will be discussed in Chapter 6. For now we will assume that the functions f and g are C^2 – that is, twice differentiable with continuous second derivatives (this assumption is for technical reasons that will become apparent later on). Further we assume that we have

$$\left.\frac{df}{dy}\right|_y \neq 0 \qquad \left.\frac{dg}{dx}\right|_x \neq 0$$

for each $y_0 < y < y_1$ and each $x_0 < x < x_1$. This condition that the derivatives of f and g do not vanish ensures that we have monotonic increasing or decreasing twists. We now define the *strength of a twist* by considering the shallowest slopes of f and g. Thus define for F:

$$\text{if } k > 0 \quad \alpha = \inf\left\{\frac{df}{dy} : y_0 < y < y_1\right\},$$

$$\text{if } k < 0 \quad \alpha = \sup\left\{\frac{df}{dy} : y_0 < y < y_1\right\} \tag{2.1}$$

and similarly for G:

$$\text{if } l > 0 \quad \beta = \inf\left\{\frac{dg}{dx} : x_0 < x < x_1\right\},$$

$$\text{if } l < 0 \quad \beta = \sup\left\{\frac{dg}{dx} : x_0 < x < x_1\right\}. \tag{2.2}$$

We call α and β the strength of the twists F and G. Both properties defined above – the wrapping number and the strength of the twist – will be important factors in the behaviour of a linked twist map. Thus we name, as in Przytycki (1983), such a twist map F a (k, α)-twist, and such a twist map G a (l, β)-twist.

Finally the *toral linked twist map* H is defined by composing F and G:

$$H : R \to R$$

$$H = H(x, y; f, g) = G \circ F.$$

It is remarkable that, depending only on inequalities involving k, l, α, and β, H can be shown to have extremely strong mixing properties on *all* of R (except for a possible set of zero measure). What we mean by 'strong mixing' is defined more precisely in Chapter 3. For the moment suffice it to say that ergodicity with positive Lyapunov exponents is a central feature of our definition.

2.2.1 Geometry of mixing for toral LTMs

In this section we give a heuristic description of how the mechanism behind a toral linked twist map relates to ideas of fluid mechanical mixing. Making this picture mathematically rigorous forms the core of much of this book.

The essence of fluid mixing via chaotic advection is stretching and folding (Ottino (1989a)). This is now a familiar concept, and in practice such stretching and folding is often created by the means of creating a fluid flow with intersecting streamlines. Of course streamlines, being defined as curves everywhere tangent to the velocity of the flow, cannot cross in a given flow pattern, so in applications flow patterns are frequently superimposed (either by introducing time dependence or spatial dependence) so that streamlines passing through a given point are at different times transverse to each other (see, for example, the mechanisms described in this and the previous chapter. In particular, this is the mechanism behind a flow commonly regarded as a paradigm example for chaotic mixing – the Aref blinking vortex (Aref (1984)) described in Section 1.5.1). This technique is now ubiquitous in mixing fluids with negligible diffusivity effects; so much so that the idea of *crossing streamlines* has become a mantra of fluid mixing.

In fact the link with dynamical systems is perhaps more directly found in *pathlines* or *streaklines*. Recall that a pathline is defined as the curve followed by a single tracer particle released into the flow, while a streakline is the curve formed by particles released into the flow at the same point but different times. For a steady flow, streamlines, pathlines and streaklines all coincide. These give the direction in which a single particle will travel, and also the direction in which a blob will align itself under the action of the flow. Thus the direction of flow along the streamlines of a steady flow give the 'characteristic directions' for the flow. In dynamical systems, providing the system has a special structure called *hyperbolicity*, the corresponding characteristic directions are called *unstable manifolds*. These will be defined rigorously in Chapter 5; roughly, the unstable manifold of a stationary point is the set of points which tend asymptotically (at an exponential rate) to the stationary point under backward evolution of the dynamical system. There is a corresponding definition for the stable manifold of a stationary point under forward evolution. Although unstable manifolds are defined under *backward* evolution, they represent the characteristic directions for *forward* evolution, in that blobs of points under forward iteration align themselves along unstable manifolds, analogously to blobs of fluid aligning themselves along streamlines. Moreover, an important result in dynamical systems (the *stable manifold theorem*) extends these ideas from stationary points to points which move around the domain (see Chapter 5 for more details). For more details on the link between unstable manifolds and fluid flow, see Beigie *et al.* (1994) for a thorough and graphic account of the way unstable manifolds form dominant structures in flows governed by chaotic advection. For now we think of equating 'unstable manifolds' with 'portions of streamlines'.

The definition of stable and unstable manifolds is such that an unstable manifold may not cross another unstable manifold, or itself (and similarly for stable manifolds). However, an unstable manifold may cross a stable manifold, and this can generate great complexity in the dynamical system (see for example Ott (1993) or Wiggins (2003)) and, in particular, produce mixing. The mechanism for such a crossing to occur comes from the fact that, typically, unstable manifolds grow in length under forward iteration, whilst stable manifolds grow in length under backward iteration. For an intersection to occur, we require the stable and unstable manifolds, whilst growing, to go, roughly speaking, in different directions (if they were parallel they would never cross). Thus the corresponding mantra for achieving enough complexity in a dynamical system so that mixing might occur is *transversely intersecting stable and unstable manifolds*.

Linked twist maps on both the torus and the plane provide a paradigm model for this to occur. Moreover, they provide a framework for rigorously:

1. establishing the existence of stable and unstable manifolds almost everywhere[1] – this gives the dynamical system the correct structure for discussing stretching and contracting;
2. demonstrating that the stable and unstable manifolds grow in length under the action of the dynamical system, and moreover showing that the lengthening stable and unstable manifolds eventually intersect each other – this results in complicated dynamical behaviour, and the fact that this conclusion is valid for almost every point in the domain means that islands cannot occur;
3. producing cases for which the the above two features are not guaranteed – this gives conditions where mixing may fail.

In the following we give a heuristic description of the relationship between linked twist maps and the kinematics of mixing in fluid flows. It is intended to clarify the key aspects of the linked twist map relevant to mixing, and to provide insight into the properties the mathematical work in later chapters will firmly establish.

---

[1] the phrase 'almost everywhere' has a technical definition which is defined and discussed in detail in Chapter 3.

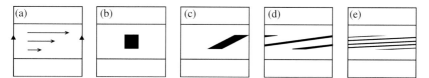

Figure 2.2 An idealized steady shear flow in a horizontal annulus. Streamlines are given in (a), and an initial blob of tracer particles in (b). Under the action of the flow the blob is sheared into the images in (c), (d) and then (e). The sheared blob becomes aligned along the direction of the streamlines, in this case horizontally.

Shear flows – toral twist maps

Imagine an idealized steady shear flow whose streamlines, pathlines and streaklines are represented by the arrows in Figure 2.2(a), where the left and right edges of the square are identified (suspending for the time being concerns about how such a flow might be created). The length of the arrows indicate that in unit time a point in the flow nearer the top of the annulus moves further than a point nearer the bottom. A blob of fluid, shown in Figure 2.2(b), is released into the flow (it may seem unlikely to have a perfectly square blob of fluid, but it helps to visualize what happens to vertical and horizontal lines). Under the action of the flow it is deformed into the images in Figure 2.2(c), then Figure 2.2(d) and finally Figure 2.2(e). The longer the flow is run (or equivalently, the stronger the effect of the shear), the closer the alignment of the points in the blob gets to the direction of the streamlines. This direction can be regarded as the most important for this shear, in that blobs are 'attracted' to it. This flow lacks the hyperbolic structure necessary for this direction to be an unstable manifold (a key difference is that nearby trajectories are attracted to this direction only *algebraically* fast, rather than *exponentially fast*), but nevertheless the horizontal direction can be regarded as the characteristic direction for this shear flow.

Figure 2.3 shows an analogous shear flow in a vertical annulus, with top and bottom edges identified. The shear is stronger at the right edge of the square than the left edge. Clearly, the characteristic direction is a different one to that for the horizontal shear. The two directions are *transverse*, and this is a crucial factor in providing good mixing.

Superimposed shear flows

The Arnold Cat Map To create a flow with 'crossing streamlines' we need to alternate the two flow patterns above. To do so we consider a flow on a torus, so that both left and right sides, and top and bottom sides, of the square are identified. Initially we widen the horizontal and vertical annuli until they are the width of the whole torus (so the domain of each shear is the same). Then

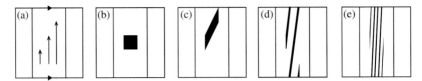

Figure 2.3 An idealized steady shear flow in a vertical annulus. Streamlines are given in (a), and an initial blob of tracer particles in (b). Under the action of the flow the blob is sheared into the images in (c), (d) and then (e). The sheared blob becomes aligned along the direction of the streamlines, in this case vertically.

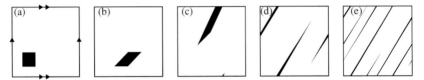

Figure 2.4 Idealized superimposed shear flows in a torus, alternating the horizontal and vertical streamlines in Figures 2.2 and 2.3, where the annuli are widened to the entire torus. An initial blob of tracer particles is shown in (a). Under the action of the flow the blob is sheared first horizontally into the image in (b), and then vertically into the image in (c). Continuing the process of alternating horizontal and vertical shears produces the images in (d) and then (e). The resulting characteristic direction is the direction of the unstable manifolds of the underlying map.

we create a flow by alternating horizontal and vertical shears. The fact that the flow directions are orthogonal means that the streamline crossing is in some sense the 'most transverse'. The resulting dynamical system is a famous and well-studied one, and one which we will refer to as a paradigm example in later chapters, and in particular define fully in Section 5.2.1. The map is known as the *Arnold Cat Map* (Arnold & Avez (1968)), and forms a fundamental example of a dynamical system for which rigorous mathematical results about mixing can be proved, and also has been used as a prototype for mixing in many applications (see for example, Childress & Gilbert (1995), Thiffeault & Childress (2003) or Boyland *et al.* (2000)).

The effect of alternating horizontal and vertical twists on a blob of fluid is shown in Figure 2.4. The initial blob becomes stretched out, again aligning itself with the characteristic direction for the system. A remarkable feature of this system is that the combination of two integrable twist maps results in a system with a hyperbolic structure, and this new direction is the direction of the unstable manifolds for the system. In general an unstable manifold need not lie along a straight line, and need not be the same for every point. However, for this simple example, the unstable manifold at every point is a straight line pointing in

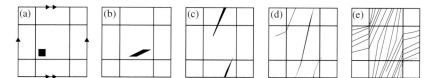

Figure 2.5 Idealized superimposed shear flows on linked annuli, alternating the horizontal and vertical streamlines in Figures 2.2 and 2.3. An initial blob of tracer particles is shown in (a). Under the action of the flow the blob is sheared into the images in (b), (c), (d) and then (e). The shearing effect only acts on the portions of the blob which are within the relevant annulus. This results in segments of the unstable manifold having differing gradients. Note however that the directions of all the line segments lie in the same quadrant.

the same direction. Moreover, this direction has irrational slope, and so, under forward evolution, the unstable manifold grows and wraps around the torus indefinitely. If we ran the system backwards we would find similar a line for stable manifolds, lying in a transverse direction. The fact that these manifolds grow and wrap around the torus makes a transverse intersection inevitable.

Toral linked twist maps When forming a linked twist map, we again alternate the action of the horizontal shear with the action of the vertical shear, but now with the original 'narrow' annuli. This leads to portions of fluid in an annulus being sheared, while other portions of fluid which lie outside the annulus are left unsheared. This system shares some of the hyperbolic structure of the Cat Map. The result of acting with this flow on an initial blob is shown in Figure 2.5. Here the unstable manifold formed is not simply a straight line. It is however a connected line consisting of line segments. These line segments are not arranged randomly, but lie in directions whose slopes may vary between horizontal and vertical. Crucially, however, all the slopes of the line segments (for this choice of system) are positive, and so just as for the Cat Map, the lengthening (under forward evolution) unstable manifold must wrap around the torus. In the next section we discuss how this property may fail to occur.

Relative directions of the toral twists A key feature (indeed, one of the most important features of linked twist maps) is the direction of the twists or shears in relation to each other. For example, in the shears illustrated and discussed above, the horizontal shear acts 'left-to-right' and the vertical shear acts 'bottom-to-top'. In other words, the shear increases with increasing coordinate for both twists. Since the coordinates on the torus can be thought of as angles we call this case *co-rotating*.

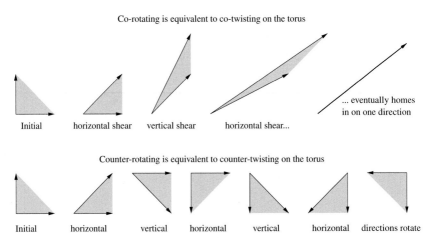

Figure 2.6 In the top figure we show the evolution of initial horizontal and vertical arrows under alternate application of horizontal and vertical shears, in the co-rotating/co-twisting case (in which both shears act in the same sense). The angle between arrows decreases, until the direction of the unstable manifold is reached, and the combined effect of the twists is to stretch areas in this direction. In the bottom figure we show the evolution of the same initial arrows under the action of a counter-rotating/counter-twisting system (in which shears act in an opposite sense). Here the arrows do not grow in length and approach each other. No overall expansion occurs because any stretching achieved by the horizontal shear is undone by the vertical shear.

Figure 2.6(a) demonstrates how the final direction of the unstable manifold is obtained. Beginning with the initial triangle on the left (we use this triangle as the bottom and left edges, marked as arrows, are aligned in the coordinate directions) we first see the result of the horizontal shear (ignoring the fact that the triangle will also be translated to a different place in the torus). The horizontal shear has no effect on the horizontal side of the triangle, but the vertical side is, of course, sheared, and the angle between arrows is decreased. The vertical shear that follows decreases the angle between arrows further (whilst the area of the triangle remains constant). Continuing the process we see the arrows lengthen and the sector formed by the arrows eventually closes to a single direction. Because the twists complement each other in this way, and the effect of the twists is to stretch in the same sense, we call this system *co-twisting*. Note that for the linked twist map on the torus, a co-rotating system is also co-twisting.

Consider now the case where we reverse the direction of the vertical shear, so that it shears from 'top-to-bottom'. We call this the *counter-rotating* case, as now one twist acts with increasing angle, and the other acts with decreasing angle. Figure 2.6(b) illustrates the evolution of the same initial triangle. After the first horizontal shear, the effect of the vertical shear is to shear downwards. As we continue applying horizontal and vertical shears, we find that we do not home in on one direction, but rather that the directions indicated by the arrows rotate. This inhibits mixing, and indeed it is not difficult to find an example of a counter-rotating system which, after six iterations of horizontal and vertical shears, returns any initial blob to its original state. The combined effect of the shears is that stretching achieved by the horizontal shear is counter acted by the vertical shear, so the system is effectively beginning to mix, and then unmixing, repeatedly. We call this *counter-twisting*. In order for a counter-twisting linked twist map to produce the necessary expansion and contraction, the shears must be sufficiently strong — stronger than for the co-twisting case.

Having established the existence of stable and unstable manifolds, and growth of their images, the final task is to show that these images intersect. For the Arnold Cat Map this was straightforward, as the images were straight lines which inevitably wrapped around the torus. In the case of the linked twist map, when parts of images of the initial blob fall outside an annulus, the effect of a horizontal or vertical shear may be omitted. Consequently, the image of an unstable manifold is a segmented line, in which some of the line segments may have different slopes (this effect can be seen in Figures 2.5(d) and 2.5(e)).

In the co-twisting case, these slopes are all constrained to fall in the same quadrant. Figure 2.7(a) shows a sketch of this. After a horizontal shear, the resulting image lies in the quadrant shown, and after a vertical shear, the same quadrant applies. Thus a simple geometrical argument shows that however line segments with such gradients are combined, the result is a segmented line which wraps around the torus.

For the counter-twisting case, shown in Figure 2.7(b), however, the shears run in opposite senses, and while after a horizontal shear, the resulting image lies in the positive quadrant as before, after a vertical shear, the directions lie in a different quadrant. When such line segments are combined, it may be possible that the resulting segmented line, although lengthening, is constrained within an enclosed region, as in Figure 2.7(b), and fail to wrap around the torus. This may prevent an intersection of stable and unstable manifolds. Chapter 7 provides an argument to say that for sufficiently strong twists, this issue does not arise.

Key point: For linked twist maps on the torus, a co-rotating system is also co-twisting, and simple conditions guarantee mixing behaviour. A counter-rotating system is also counter-twisting, and in this case stronger conditions on the strength of the twists are required to guarantee mixing.

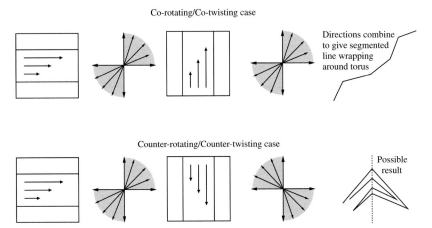

Figure 2.7 Summary of the effect of the sense of the shears in a linked twist map on the unstable manifold. For the co-rotating case the shears run in the same sense, and the unstable manifold is constructed of line segments all with positive gradient. For the counter-rotating case, shears run in opposite senses, and the unconstrained line segments may cause the unstable manifold to double back on itself and fail to wrap around the torus.

2.3 Linked twist maps on the plane

The relationship between toral linked twist maps and fluid flows is probably far from obvious. The first step in understanding this relationship is by considering how linked twist maps on a subset of the plane can arise in fluid flows (and then later consider the relationship between toral linked twist maps and linked twist maps on subsets of the plane).

Consider a region of fluid, possibly a 2D cavity with solid boundaries or the cross flow in a 3D steady flow in a channel with an axial flow in the direction normal to the page, containing a stagnation point surrounded by closed streamlines, as shown in the top panel of Figure 2.8.

The fact that we are showing our region to be a (enclosed) square with circular streamlines is unimportant for our conclusions. The horizontal and vertical dashed lines are axes centred in the middle of the region and merely serve as a (essentially arbitrary) reference point. The stagnation point is on the horizontal axis, offset to the left of the centre. Imagine that at some later time, and by some mechanism (the details of which are unimportant to the argument) the flow pattern is altered. The alteration involves moving the stagnation point to the right of the centre, as shown in the lower panel of Figure 2.8. We remark that the flow domain could contain multiple 'recirculation cells', and our arguments can be applied to any number of them. The flow cycles between the upper and lower

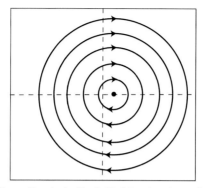

Streamlines in the first half of the advection cycle

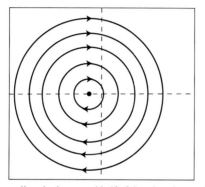

Streamlines in the second half of the advection cycle

Figure 2.8 Geometry of the flow patterns for the two half cycles of the advection cycle.

patterns in a periodic fashion, which is the advection cycle of interest to us. The cycling could occur as a result of the imposition of periodic time dependence in a cavity flow, or as a result of the flow pattern changing periodically in the cross section of a flow as in a discontinuous duct flow. The utility of this approach is that it is independent of the details of how the flow is created, and depends only on the geometry of the flow patterns. Clearly also the length of the period of cycling will play an important role in the rate of mixing, and we will address this later.

The fluid particle motion from the beginning of a half cycle to the end of the same half cycle is described by a twist map. For the closed streamlines in each half cycle let (r, θ) denote streamline coordinates (the actual trajectories in the x–y plane need not be circular; they can be made circular by some nonlinear transformation). That is, on a streamline r is constant and θ is an angular variable that increases monotonically in time. The map of particles from the beginning

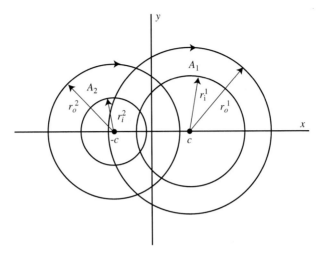

Figure 2.9 Geometry of the annuli from each half cycle of the advection cycle that make up the linked twist map.

to the end of a *half* cycle is given by $S(r, \theta) = (r, \theta + g(r))$. We will see that the function $g(r)$ is the key here. It provides the angular displacement along the streamline during one half of the advection cycle. Typically it varies from streamline to streamline, and it is from this property that the phrase 'twist map' arises. For example, $g(r)$ may increase as r moves from the elliptic stagnation point to a particular streamline (i.e., $g(r)$ achieves a unique maximum), and then it decreases monotonically to zero on the boundary.

We can now define a linked twist map over a complete advection cycle. Let A_1 denote an annulus whose inner (denoted r_i^1) and outer (denoted r_o^1) boundaries are streamlines centred at $(c, 0)$ in the first half of the advection cycle. Let A_2 denote an annulus constructed in the same way centred at $(-c, 0)$ in the second half of the advection cycle. We show the two annuli in Figure 2.9. The two annuli must be chosen so that they intersect each other transversally in the sense that they intersect each other in two disjoint regions, as shown in the figure.

An obvious question arises: How do we choose the annuli A_1 and A_2? Or, which annuli A_1 and A_2 do we choose? Since there are clearly many such choices of pairs of annuli that will satisfy the transversal intersection condition. The theory of linked twist maps does not answer this question. It is only concerned with the behaviour on a pair of annuli satisfying the transversal intersection property, and some additional properties described below. For mixing purposes, in a given situation we will need to find the largest and/or largest number of annuli in the flow domain satisfying the hypotheses. We will next discuss these hypotheses on a given pair of annuli. In each half cycle depicted in Figure 2.8 we are only showing one recirculation region, i.e., one region of closed streamlines. In some applications a half cycle may contain multiple recirculation regions (separated by heteroclinic and/or homoclinic orbits). This poses no difficulty in the application of the linked twist map approach since we merely need to find transversally intersecting annuli in each half cycle. However, it does raise questions about how many annuli can be found and, as we shall see, the sense of rotation may also be important (i.e., the issue of co-rotation versus counter-rotation as discussed in the previous section).

Let $S_1(r,\theta) = (r,\theta + g_1(r))$ be a twist map defined on A_1 with $dg_1/dr \neq 0$ and $g_1(r_i^1) = 2\pi n$, for some integer n. Then r_o^1 is chosen such that $g_1(r_o^1) = 2\pi(n + k_1)$, where k_1 is a nonzero integer. Let $S_2(r,\theta) = (r,\theta + g_2(r))$ be a twist map defined on A_2 with and $g_2(r_i^2) = 2\pi m$, for some integer m, and $g_2(r_o^2) = 2\pi(m + k_2)$, for some nonzero integer k_2. Furthermore, we suppose that the annuli intersect transversally in two disjoint components in the sense that $r_i^1 \cap r_i^2 \neq \emptyset$ and $r_o^1 \cap r_o^2 \neq \emptyset$, as shown in Figure 2.9. The map defined by $S_2 \circ S_1$ on $A_1 \cup A_2$ is referred to as a linked twist map on a subset of the plane.[2]

Depending on the signs of k_1 and k_2, the strengths of the twists can be defined in the same manner as in (2.1) and (2.2). Now the relationship between toral linked twist maps and linked twist maps defined on a subset of the place should be clear in the sense that each is created by defining a twist map on an annulus, the annuli intersect (and some care must be taken here in the planar case), and the linked twist map is the composition of two twist maps defined on each annulus. However, the transference of the strong mixing results obtained for toral linked twist maps to linked twist maps defined on a subset of the plane is geometrically more complicated, but it can be done (one gets an idea of the subtleties that may arise when one considers the fact that the two annuli on the torus intersect in one component and the two annuli in the plane intersect in two components).

We will see that for designing a flow with optimal mixing the key quantities we need to understand are $g_i(r)$, $i = 1, 2$, on the annuli of choice, since these functions determines the rotation properties of the annuli, the radii of the

---

[2] As discussed in more detail for the case of toral linked twist maps, strictly speaking, the map $S_2 \circ S_1$ is not defined on all of $A_1 \cup A_2$ since S_1 is only defined on A_1 and S_2 is only defined on A_2. In this situation we can define S_1 (resp., S_2) to be the identity map on the part of $A_1 \cup A_2$ that is not A_1 (resp., the part of $A_1 \cup A_2$ that is not A_2). Since S_1 (resp., S_2) is the identity map on the boundary circles defining A_1 (resp., A_2) this extension of the maps to a larger domain can be done smoothly. Conceptually, from the physical point of view, this point can be ignored. It says nothing more than we consider S_1 (resp., S_2) only acting on A_1 (resp., A_2), and it does nothing to points that are in the part of A_2 (resp., A_1) that does not intersect A_1 (resp., A_2).

annuli, and the strength of the twist, which in the context of fluids, is the shear rate. This is significant because the design of a mixer with strong mixing properties effectively reduces to the properties of one function describing closed streamlines in each half cycle of the advection cycle.

Finally, there is a technical problem that must be addressed in applying the currently known LTM results for the purpose of concluding that the mixing has the Bernoulli property on the two chosen annuli. Let us describe the situation above a bit more carefully. We are concerned with two flow patterns that alternate (blinking flow); call them pattern 1 and pattern 2. Now the LTM formalism and results as developed by mathematicians do not focus on all of pattern 1 and pattern 2. Rather, they focus on one annulus (in isolation) in pattern 1 and one annulus (in isolation) in pattern 2 that intersect 'transversely'.

Hence, in the mathematician's formalism, the LTM maps particles between the annuli as we alternate applying the twist maps to each annulus, and the theorems describe mixing of particles in the two annuli. Now in order for this to make sense, the *same* particles must remain in the two annuli for all time. This is not true for two arbitrarily chosen flow patterns. However, it is true if pattern 2 is a rigid rotation of pattern 1 (this is precisely what is done in the analysis of the partitioned pipe mixer described in Khakhar *et al.* (1987)). Since the flow is bounded, and we know where all the particles go, we suspect that some strong mixing results can be proven in the case where this is not true. However, this is a problem that awaits further mathematical analysis. Further, we remark that if one is only interested in constructing Smale horseshoes we do not require this condition since the invariant set associated with the horseshoe is directly constructed in the overlap regions between two appropriately chosen annuli (see Devaney (1978), Wiggins (1999), Khakhar *et al.* (1986), and Chapter 4 where Smale horseshoes are discussed in detail).

So LTMs on subsets of the plane mix 'well', where 'well' can be made mathematically precise. We will make a more explicit connection between LTMs on a subset of the plane and the geometry of mixers with a number of examples shortly. However first we give a discussion of the heuristics of fluid mechanical mixing in planar linked twist maps analogously to that given in Section 2.2.1.

2.3.1 Geometry of mixing for LTMs on the plane

Briefly, precisely the same mechanism described in Section 2.2.1 is at work for the case of linked twist maps on the plane, although here the curvilinear coordinate system makes the analysis slightly more complicated. We have a pair of shears again, each one taking place on an annulus in the plane. The two annuli overlap each other, creating a pair of intersections on which points may be acted

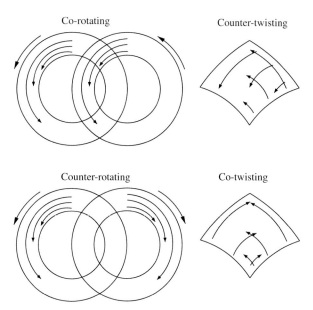

Figure 2.10 On the plane, a pair of co-rotating shears produce a counter-twisting effect in the intersection, while a pair of counter-rotating shears produce a co-rotating effect.

on by both shears. As for the toral case, we are interested in seeing the effect of the shears on unstable manifolds. Again we would like the line segments to lie in directions such that the lengthening unstable manifold will wrap around an annulus rather than double back on itself. The same dichotomy presents itself – that of co-rotating and counter-rotating systems. To be consistent with the majority of the literature on shear flows, we define a co-rotating system to be a system for which the shears either both act clockwise, or both act counter-clockwise (that is, both with increasing angle or both with decreasing angle). Note that this definition is the same as for toral linked twist maps. Similarly, if the shears act in opposite senses, the system is a counter-rotating linked twist map.

When considering the relative directions of the twists, however, care must be exercised. The complexity in the dynamics takes place in the intersections of the annuli – it is here that we have stretching in transverse directions. Consequently we must inspect the behaviour of the shears within the intersections. In fact the intersections of annuli on the place are locally the same as the intersections of annuli on the torus.

Figure 2.10 illustrates the relationship between the direction of rotation of the shears on the planar annuli to the direction of the shears within the intersection of the annuli. In Figure 2.10(a) we have a co-rotating system, in which the shears both act in a counter-clockwise sense. However, inspecting the behaviour of the shears within the upper intersection reveals that the shears are acting in a counter-twisting manner (precisely the same conclusion can be deduced for the lower intersection). On the other hand, if the planar shears run in a counter-rotating sense (as in Figure 2.10(b)), the behaviour in the intersection produces a co-twisting linked twist map.

As for the toral linked twist maps, stronger conditions are necessary to establish results for the counter-twisting (co-rotating) system on the plane then for the co-twisting (counter-rotating) system.

Key point: A co-rotating linked twist map on the plane is a counter-twisting linked twist map, while a counter-rotating linked twist map on the plane is a co-rotating linked twist map.

2.4 Constructing a LTM from a blinking flow

As we have mentioned, one of the first flows that was demonstrated to possess chaotic advection was the *blinking vortex flow* (Aref (1984)). The idea is simple. Consider a streamfunction $\psi_A(x, y)$ that defines a particular flow pattern (i.e., streamline structure), and another streamfunction, $\psi_B(x, y)$, that defines a different flow pattern. Imagine that we allow particles to evolve under the flow defined by $\psi_A(x, y)$ for a certain time interval, then we 'stop' the flow defined by $\psi_A(x, y)$ and then let the particles evolve under the flow defined by $\psi_B(x, y)$ for some time interval. At the end of this time interval we 'stop' the flow generated by $\psi_B(x, y)$, and then repeat this cycle indefinitely.

Mathematically, we can make this more precise. Consider the streamfunction:

$$\psi(x, y, t) = f_A(t)\psi_A(x, y) + f_B(t)\psi_B(x, y),$$

where

$$f_A(t) = \begin{cases} 1 & kT < t \leq kT + \frac{T}{2} \\ 0 & kT + \frac{T}{2} < t \leq (k+1)T \end{cases}$$

and

$$f_B(t) = \begin{cases} 0 & kT < t \leq kT + \frac{T}{2} \\ 1 & kT + \frac{T}{2} < t \leq (k+1)T \end{cases}$$

Then the velocity field is given by:

$$\dot{x} = \frac{\partial \psi}{\partial y},$$

$$\dot{y} = -\frac{\partial \psi}{\partial x}.$$

One sees from the form of the functions $f_A(t)$ and $f_B(t)$ that the flow 'blinks' periodically between the (steady) flow defined by $\psi_A(x, y)$ and the flow defined by $\psi_B(x, y)$. Following the discussion in Section 2.3, we need to select annuli defined by streamlines of $\psi_A(x, y)$ and annuli defined by streamlines of $\psi_B(x, y)$ having the property that these two annuli 'link' in the manner described in the discussion in Section 2.3. Of course, these annuli must also be chosen to satisfy the conditions of the LTM theorems in order to guarantee good mixing. However, the great variety of streamline topologies gives rise to many possible choices for annuli, as well as pointing the way to many new areas of research in extending the LTM framework. In Chapter 1 we gave some examples which can be modelled by blinking flows. These also indicate new areas of research for extending the LTM framework.

2.5 Constructing a LTM from a duct flow

Now we show how to rigorously obtain a linked twist map (LTM) from a sequence of duct flows using results in Mezic & Wiggins (1994). Streamline crossing occurred at different times in the blinking flows that we considered earlier. In the duct flows 'streamline crossing' occurs as a result of spatial periodicity in periodically displaced cross-sections of the flow along the duct.

Recall that by a *duct flow* we mean a three-dimensional, steady, spatially periodic flow where the axial flow does not depend on the axial coordinate, as defined earlier. We refer to each spatial period as a *cell*, and the flow is described by a mapping from the beginning of a cell to the end of a cell. This mapping is the composition of two mappings, each a twist map (which is how the LTM arises). The first twist map is the mapping of particles in the cross-section at the beginning of the cell to the cross-section at the half cell. The second twist map is the mapping of particles from the cross-section at end of the half cell to the cross-section end of the cell.[3] The cell to cell LTM obtained in this way

---

[3] There is no reason that a cell should be divided into two 'half cells'. It could be divided into three, four, or *n* different sub-cells. In principle, the analysis can be carried out by the same approach, but it would require a generalization of the classical LTM approach.

will be discontinuous in the sense that each twist map is computed separately and continuity at the half cell is not enforced.

Construction of the first twist map

We assume that we have a duct flow in the first half of the cell having the following form:

$$\frac{\mathrm{d}x}{\mathrm{d}t} = \frac{\partial \psi_1(x, y)}{\partial y},$$

$$\frac{\mathrm{d}y}{\mathrm{d}t} = -\frac{\partial \psi_1(x, y)}{\partial x}, \tag{2.3}$$

$$\frac{\mathrm{d}z}{\mathrm{d}t} = k_1(x, y).$$

Since the x and y components of the velocity field (i.e., the cross flow) *do not* depend on the axial coordinate z we can consider transforming this two-dimensional velocity field into standard action-angle variables. To do this, we must make the following assumption.

Assumption There is some connected subset of the x–y plane, denoted $\mathcal{D}$, in which the level sets $\psi_1(x, y) = c$ are closed 'streamlines'.

Action-angle transformation in the cross flow If this assumption holds, then it is well known from classical mechanics (see, e.g. Arnold (1978)) in this region there is a transformation

$$(x, y) \mapsto (r, \theta)$$

satisfying the following properties.

1. $r = r(c)$, i.e., r is constant on the closed streamlines in the cross flow.
2. $\oint_{\psi_1=c} \mathrm{d}\theta = 2\pi$.
3. $\dot{\theta} = \Omega_1(r)$

The action variable is given by (see e.g. Wiggins (2003))

$$r = \frac{1}{2\pi} \int_{\psi_1=c} y \mathrm{d}x,$$

while the angle variable is given by:

$$\theta = \frac{2\pi}{T(\psi_1)} t,$$

where $T(\psi_1)$ is a period on the orbit in the cross flow (which is a level set of ψ_1), and t denotes the time along the streamline measured from a certain starting point on the streamline.

We assume that this action-angle transformation on the x–y component of (2.3) has been carried out so that these equations subsequently take the form

$$\dot{r} = 0,$$
$$\dot{\theta} = \Omega_1(r), \qquad (2.4)$$
$$\dot{z} = h_1(r, \theta),$$

where

$$h_1(r, \theta) = k_1(x(r, \theta), y(r, \theta)).$$

We remark that there may be multiple regions of closed streamlines in the cross flow separated by homoclinic or heteroclinic orbits, or even solid boundaries. In general, a separate action-angle transformation is required for each such region of closed streamlines (and the different transformations may not be simply related).

Action-angle-axial transformation in the half cell

Now we introduce a final change of coordinates that leaves the action-angle variables in the cross flow unchanged, but modifies the axial coordinate so that all particles on a given streamline in the cross flow take the same time to travel the length of the half cell. We refer to these coordinates as *action-angle-axial* coordinates (Mezic & Wiggins (1994)).

Suppose $\Omega_1 \neq 0$ in (2.4). Then the transformation of variables

$$(r, \theta, z) \rightarrow (r, \theta, a),$$

defined by

$$r = r$$
$$\theta = \theta,$$
$$a = z + \frac{\Delta z_1(r)}{2\pi} \theta - \int \frac{h_1(r, \theta)}{\Omega_1(r)} d\theta,$$

where

$$\Delta z_1(r) = \frac{1}{\Omega_1(r)} \int_0^{2\pi} h_1(r, \theta) d\theta,$$

then brings the system (2.4) to the form

$$\dot{r} = 0,$$
$$\dot{\theta} = \Omega_1(r),$$
$$\dot{a} = A_1(r),$$

where

$$A_1(r) = \frac{\Delta z_1(r)}{2\pi} \Omega_1(r). \tag{2.5}$$

Furthermore, the transformation is volume-preserving.

Explicit expression for the twist map Now a linked twist map can be constructed from the velocity field in (r, θ, a) coordinates. Let $z = 0$ be the starting point of the channel. From the coordinate transformation above we see that the beginning of the first half cell, $z = 0$, corresponds to:

$$a = \frac{\Delta z_1(r)}{2\pi}\theta - \int \frac{h_1(r,\theta)}{\Omega_1(r)} d\theta.$$

The end of the first half cell (beginning of the second half cell) is at $z = \frac{L}{2}$, which corresponds to:

$$a = \frac{L}{2} + \frac{\Delta z_1(r)}{2\pi}\theta - \int \frac{h_1(r,\theta)}{\Omega_1(r)} d\theta.$$

The time of flight, T, from the beginning of the first half cell to the end is given by solving:

$$a(T) = A_1(r)T + a(0),$$

where

$$a(0) = \frac{\Delta z_1(r)}{2\pi}\theta - \int \frac{h_1(r,\theta)}{\Omega_1(r)} d\theta.$$

and

$$a(T) = \frac{L}{2} + \frac{\Delta z_1(r)}{2\pi}\theta - \int \frac{h_1(r,\theta)}{\Omega_1(r)} d\theta.$$

After some simple algebra we easily find that $T = L/2A_1(r)$. Therefore the twist map for the first half cell is given by:

$$r \mapsto r,$$

$$\theta \mapsto \theta + \frac{L\Omega_1(r)}{2A_1(r)}$$

or, using (2.5),

$$r \mapsto r,$$

$$\theta \mapsto \theta + \frac{\pi L}{\Delta z_1(r)}.$$

Construction of the second twist map

The construction of the twist map in the second half of the cell proceeds in exactly the same way as the construction for the first half. We suppose the flow in the second half is given by:

$$\frac{dx}{dt} = \frac{\partial \psi_2(x, y)}{\partial y},$$
$$\frac{dy}{dt} = -\frac{\partial \psi_2(x, y)}{\partial x},$$
$$\frac{dz}{dt} = k_2(x, y),$$

where $\psi_2(x, y)$ describes the streamline pattern in the cross flow, and $k_2(x, y)$ describes the axial velocity. We then proceed with the same series of transformations as above.

2.6 More examples of mixers that can be analysed in the LTM framework

All of the examples discussed in Chapter 1 are amenable to analysis via the linked twist map approach. In this section we describe more systems that fit within the framework. This is significant because, as we will see in the following chapters, LTMs provide an analytical approach to the design of devices that produce a mathematically optimal, and precisely defined, type of mixing.

We point out again that the first example of a chaotic flow, the blinking vortex flow (Aref (1984)), is also the most transparent and the most immediately analysable example. In this case the flow itself is already in the form of a linked twist map on the plane and the appropriate functions can be controlled at will. This connection was described in Wiggins (1999). It is remarkable that, in some sense, this example encompasses, if not all, a large number of other examples. The egg beater flows also have this 'universal' characteristic, and are examples of linked twist maps on the torus. It is also important to stress that the most conceptually efficient way to think about mixing is in terms of maps and not in terms of deviations from integrability. Deviations from integrability, by definition, lead to confined, as opposed to widespread, mixing. The most useful heuristic is streamline crossing, i.e. streamlines in a bounded domain at two different times must intersect. This is precisely the central message, and mechanism, of the LTMs.

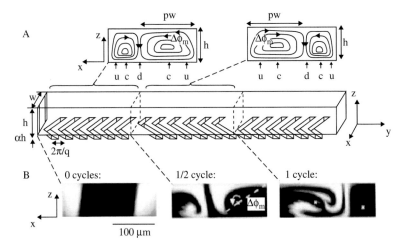

Figure 2.11 Figure 2 from Stroock *et al.* (2002) showing the closed streamlines in the cross-section of each half-cycle. Panel A shows the mixer and panel B shows flow visualizations in the cross-section.

We consider a few examples, the first from recent devices intended for micro-fluidic applications, others from older systems that illustrate mechanisms that may be used in future microfluidic applications.

One possibility for flow manipulation is the use of patterned walls. Stroock *et al.* (2002) built and conducted experiments in a micromixer consisting of a straight channel with ridges placed on one of the walls of the channel at an oblique angle with respect to the axis of the channel. When the fluid is driven axially by a pressure gradient the ridges on the floor of the channel give rise to a transverse flow. In the *x–y* plane or cross flow the streamlines are closed and helical in three dimensions. If the ridges are arranged in a periodic pattern down the axis of the channel, a herringbone pattern zigzagging to the right and to the left, each period consists of two half-cycles, producing two cells. If the pattern is such that the two cells are asymmetric with respect to the *y*-axis, and one looks at the mixer along the axial path, then the elliptic points corresponding to the centre of the cells switch positions after one cycle. The overall map then consists of the composition of two maps: the maps between each half-cycle. The map between cycles is a linked twist map. This is shown in Figure 2.11 from Stroock *et al.* (2002). The linked twist map results described here provide a basis for design and analysis of rigorously defined mixing properties in such flows.

Several extensions become apparent. For example, if the cross-section is not mirror symmetric as is a rectangle, but is, for example, trapezoidal, a

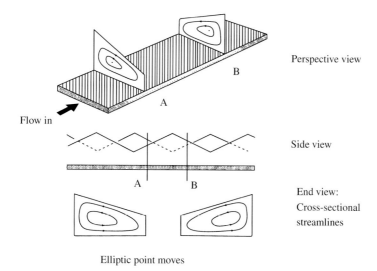

Perspective view

Side view

End view:
Cross-sectional
streamlines

Elliptic point moves

Figure 2.12 Micromixer with a periodically varying trapezoidal cross-section driven by a grooved bottom wall. A and B denote the beginning of each half cell where the cross-sectional flow is shown. [Figure from Wiggins & Ottino (2004)].

herringbone pattern is unnecessary. In this case a patterned wall is all that is needed. The key idea is to shift the location of the elliptic point (see Figure 2.12).

There are many other examples that can be fitted within the LTM framework. In fact, some of these examples are older, going back to static mixing concepts. The Partitioned Pipe Mixer (PPM) of Khakhar *et al.* (1987) and analysed experimentally by Kusch & Ottino (1992) is representative of a large class of spatially periodic flows, and the first continuous flow that was shown to be chaotic. The PPM consists of a pipe partitioned into a sequence of semi-circular ducts by means of orthogonally placed rectangular plates (Figure 2.13).

A cross-sectional motion is induced through rotation of the pipe wall. At every length L along the pipe axis, the orientation of the dividing plate shifts by $90°$. Thus a series of two co-rotational flows is followed by two co-rotational flows but shifted by $90°$. This particular geometry of the PPM is just one of three possible spatially periodic configurations which could be realized. Franjione & Ottino (1992) considered two variants of this flow. One that qualitatively captures the motion in a sequence of pipe bends (the 'twisted-pipe' flow of Jones *et al.* (1989)) – two counter-clockwise rotating vortices followed by two counter-clockwise rotating vortices; the other a flow that resembles the motion in a Kenics®static mixer – two counter-clockwise rotating vortices followed by two clockwise rotating vortices. As discussed earlier in this article the PPM design fits precisely the LTM formalism.

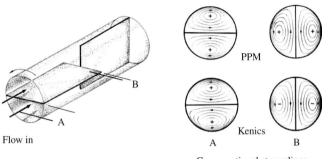

PPM

Kenics

Flow in

A   B

Cross-sectional streamlines

Figure 2.13 The partitioned pipe mixer. The left-hand panel shows the duct and the internal arrangement of plates. In the right-hand panel the top of the figure show the cross-sectional flows for the PPM at the indicated locations, and the lower part of the panel shows the cross-sectional flows that would occur for the Kenics®mixer. [Figure adapted from Franjione & Ottino (1992).]

Static mixers try to mimic the Baker's transformation (which is discussed in Chapter 3). Static mixers invariably involve internal surfaces; two regions have to be split and then reconnected (this tries to mimic cutting in 2D). This is an issue that makes these systems complicated to build at small scales. Some remarkably small mixers have been built (e.g. Bertsch *et al.* (2001)), but these designs are scaled-down versions of a design commonly encountered in routine large-scale applications. However, this need not be so and other design possibilities, more in line with current microfabrication technologies, should be explored. Nevertheless large-scale applications may be a source of inspiration. Consider the Rotated Arc Mixer (Metcalfe *et al.* 2006). The design depends critically on a clever use of the cross-sectional flow. The system consists of two hollow cylinders with a very small gap between them; the outer cylinder rotates while the flow is driven axially by a pressure gradient. The inner cylinder has strategically placed cut offs, exposing the flow contained in the inner cylinder to the drag of the moving outer cylinder (Figure 2.14).

In the example in Figure 2.14 there are two cut offs per period, but obviously the system can be generalized to any number of cut offs. With two cut offs per period the system corresponds exactly to an LTM. The theory for the case of more than two cut offs has not been yet developed. It is apparent that this design can be implemented, at least in theory, by means of electro-osmotic flows. Another variation on the cavity flow is to exploit time-dependent changes in geometry by adding a secondary baffle (Figure 2.15).

This idea goes back to Jana *et al.* (1994a), and variations on this idea have been patented in the context of polymer processing applications. In the original case the cross flow was induced by an upper wall sliding diagonally. However,

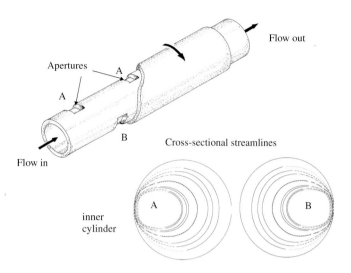

Figure 2.14 The rotated arc mixer. [Figure adapted from Metcalfe *et al.* (2006).]

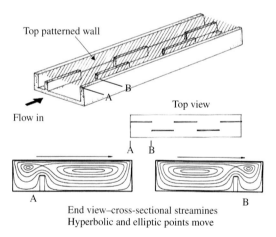

Figure 2.15 Micromixer with a periodic series of baffles driven by a grooved bottom wall. A and B denote the beginning of each half cell where the cross sectional flow is shown. [Figure from Wiggins & Ottino (2004).]

it is easy to see that the design will work if the driving is due to patterned ridges as developed by Stroock *et al.* (2002). In this case the portrait changes from a Figure-8 to another Figure-8 but where the location of the central hyperbolic point has been shifted (Jana *et al.* (1994a)). In some sense this resembles two

LTMs, but the hyperbolic point changes the mathematical structure. The theory for this case remains to be developed and the function $g(r)$ has a double hump structure. However, the same mathematical approach taken in the original LTM papers should apply.

All the above designs are spatially periodic. For example, in the case of the PPM mixer, we have a sequence of vertical and horizontal elements; in a static mixer, a sequence of right-handed (R) and left-handed (L) elements; in the Stroock mixer a series of elements zigzagging to the right and to the left. Thus, in all of the above examples, when we assemble the mixer we have a sequence of the form R–L–R–L–R, and so on. The idea is good, but it has limitations. For example there may be unmixed regions that persist even with an infinite number of elements and it is clearly possible for unmixed tubes to exit the mixer without ever mixing with the surroundings. In the past the only way to investigate the presence of islands was to propose a design and to resort to computations. However, to the extent that a sequence R–L can be viewed as a twist map, we can now be assured that mixing is effective in a region that can be calculated a priori. The task therefore is to mix the well-mixed regions with the unmixed regions. This can be accomplished by designing a suitable follow-up twist map, say R′–L′, or by symmetry manipulations (Franjione & Ottino (1992)).

3

The ergodic hierarchy

This chapter establishes the mathematical foundations on which to build. The key concepts and results from ergodic theory are given, forming an ordered list of behaviours of increasing complexity, from ergodicity, through mixing, to the Bernoulli property.

3.1 Introduction

We will mostly be interested in the simplest mixing problem: the mixing of a fluid with itself. This serves as a foundation for all mixing problems. Practically, we can think of placing a region, or 'blob' of dye in the fluid, and asking how long it takes for the dye to become evenly, or uniformly, distributed throughout the entire domain of the flow. We will need the mathematical machinery to make this question precise and quantitative. We will first need a framework to mathematically describe, measure and move regions of fluid. To do so we will introduce simple ideas and definitions from topology, measure theory and dynamical systems. In particular, notions of set theory from topology correspond naturally to properties of an arbitrary region of fluid, such as its boundary, interior and connectedness. Measure theory provides the tools necessary to measure the size of a region in a generalized and consistent way. The basis of the field of dynamical systems is the study of the evolution of some system with time, and these ideas can be applied directly to the application of moving fluid. These are by now all well-established techniques in the study of fluid mechanics (see for example Ottino (1989a)).

To discuss mixing of fluid we will need the area of mathematics known as *ergodic theory*. This provides a framework in which many physically relevant pieces of the mixing problem can be fruitfully studied. We work our way through the 'ergodic hierarchy'. This is an ordering of behaviours of increasing complexity; starting with *ergodicity*, which is a notion of indecomposability, and means that typical trajectories traverse the whole region in question; through *mixing*, which has a technical definition that formalizes the intuitive idea of mixing of fluids; and finishing with the *Bernoulli property*, which is the most complex – in fact it implies that the system is statistically equivalent to a series of random coin tosses.

From the point of view of an engineer, physicist, or applied mathematician, modern ergodic theory may appear to be a highly technical subject. Moreover, typical studies and advances frequently seem to have very little in common with 'real applications'. However, the same criticism could probably have been levelled at the dynamical systems theory of the 1960s, and yet throughout the 1980s, and continuing to the present day, dynamical systems theory has developed a collection of techniques, tools, and ways of approaching problems that have become indispensable to a wide, and diverse, range of disciplines. We believe that ergodic theory will ultimately prove to be just as useful in many different disciplines, and this monograph could be viewed as an attempt to provide a guide to aspects of 'applied ergodic theory' by showing how certain ideas from the subject can be put to immediate use in mixing problems that arise in a number of different areas. Nevertheless, the reader will still find it necessary, and useful, to consult the mathematical literature on the subject, but this could prove difficult due to the level of mathematical background required as a starting point. We will attempt to provide a bridge to the literature by giving a more heuristic description of aspects of the theory and explaining why certain, seemingly abstract, ideas are needed in the context of the problem of the mixing of a fluid with itself. Examples of excellent and thorough textbooks on ergodic theory are Arnold & Avez (1968), Halmos (1956), Keller (1998), Lasota & Mackey (1994), Mané (1987), Parry (1981), Petersen (1983), Pollicott & Yuri (1998), Sinai (1994), and Walters (1982), and many more undoubtedly exist.

3.2 Mathematical ideas for describing and quantifying the flow domain, and a 'blob' of dye in the flow

We now develop some of the mathematical tools and terminology to describe the structure and size of the flow domain and of a 'blob' of dye in the flow, using elementary notions from topology and measure theory. Both the flow domain and the blob is mathematically described by a chosen *set* of points. The more applications oriented reader may regard some of the mathematical abstraction in this section as unnecessarily distracting, and there probably is some justification for this. However, a familiarity with the mathematician's way of writing about the subject is crucial for discovering what results in the abstract ergodic theory literature might be relevant for applications. For this reason, some level of understanding of the mathematical language of ergodic theory and dynamical systems theory is essential.

We will require some familiarity with basic elements of set theory and operations on sets such as union, intersection, complement, etc., and we will address this in more detail later. First, in order to describe the relationship of blobs (sets) and points in the flow domain (mathematically, also a set) we must endow our domain with some additional mathematical structure. In particular, it will be extremely useful (essential for our purposes, in fact) to be able to describe the distance between two points. For this reason, we will require our flow domain to be a *compact metric space*, which we define in the following section.

3.2.1 Mathematical structure of spaces

Mathematical analysis is underpinned by the very general concept of a space, which formally endows a structure, or set of rules and relationships, on collections of points. In this section we give some details of the types of structured space we will meet throughout the book. In particular, in order to use familiar and fundamental mathematical ideas such as distance, length, angle, connectedness, continuity and differentiation we will require constructs such as the *metric space*, the *vector space*, the *topological space* and the *manifold*. These definitions may seem technical and far removed from applications, but it is the strength of many mathematical results that we shall quote that they hold in extremely abstract circumstances, and require little specificity. In fact, as we mention repeatedly, all the spaces we shall be concerned with are common spaces (such as Euclidean space $\mathbb{R}^n$) which naturally satisfy the necessary requirements. We give the following definitions in just enough detail to be able to discuss later results. For further information the reader should consult any good book on mathematical analysis, for example Munkres (1975), Conway (1990) or Halmos (1950).

Metric space

A metric space is a space endowed with a distance function, or metric, which defines the distance between any pair of points in the set. The metric is a non-negative function, and has intuitive, yet important, properties.

Definition 3.2.1 (Metric) *A metric $d(x, y)$ is a function defined on pairs of points x and y in a given set M, such that*

1. $d(x, y) = d(y, x)$ *(i.e., the distance between x and y is the same as the distance between y and x),*

2. $d(x, y) = 0 \iff x = y$ (*i.e., the distance between a point and itself is zero, and if the distance between two points is zero, the two points are identical*),
3. $d(x, y) + d(y, z) \geq d(x, z)$ (*the 'triangle inequality'*).

It is an instructive exercise to see that the three properties of a metric imply that $d(x, y) \geq 0$, so we never have a negative distance between two points.

Definition 3.2.2 (Metric space) *A set M possessing such a metric is called a metric space.*

The most familiar metric space is the Euclidean space, $\mathbb{R}^n$. Here the distance between two points $x = (x_1, \ldots, x_n), y = (y_1, \ldots, y_n)$ is given by the Euclidean metric $d(x, y) = \sqrt{(x_1 - y_1)^2 + \cdots + (x_n - y_n)^2}$. It is a simple task to confirm that the Euclidean distance does indeed satisfy the three properties of a metric given in Definition 3.2.1.

Vector space

A vector space V, also known as a linear space, is a set that is closed under vector addition and scalar multiplication. Elements of V are called vectors. We will not list explicitly all the properties that vectors are required to possess, but these can be found in any textbook on linear algebra (for example, Halmos (1974)). Again Euclidean space $\mathbb{R}^n$ is the standard example of a vector space, in which vectors are a list of n real numbers (coordinates), scalars are real numbers, vector addition is the familiar component-wise vector addition, and scalar multiplication is multiplication on each component in turn.

Normed vector space

To give a vector space some useful extra structure and be able to discuss the length of vectors, we endow it with a *norm*, which is closely related to the idea of a metric. A norm gives the length of each vector in V.

Definition 3.2.3 (Norm) *A norm is a function* $\| \cdot \| : V \to \mathbb{R}$ *which satisfies:*

1. $\|v\| \geq 0$ *for all* $v \in V$ *and* $\|v\| = 0$ *if and only if* $v = 0$ *(positive definiteness)*
2. $\|\lambda v\| = |\lambda| \|v\|$ *for all* $v \in V$ *and all scalars* λ
3. $\|v + w\| \leq \|v\| + \|w\|$ *for all* $v, w \in V$ *(the triangle inequality)*

It is easy to see the link between a norm and a metric. For example, the norm of a vector v can be regarded as the distance from the origin to the endpoint of v. More formally, a norm $\|\cdot\|$ gives a *metric induced by the norm* $d(u, v) = \|u - v\|$.

Whilst the Euclidean metric is the most well known, other norms and metrics are sometimes more appropriate to a particular situation. A family of norms called the L^p-norms are frequently used, and are defined as follows:

$$L^1\text{-norm} : \|x\|_1 = |x_1| + |x_2| + \cdots + |x_n|$$

$$L^2\text{-norm} : \|x\|_2 = (|x_1|^2 + |x_2|^2 + \cdots + |x_n|^2)^{1/2}$$

$$L^p\text{-norm} : \|x\|_p = (|x_1|^p + |x_2|^p + \cdots + |x_n|^p)^{1/p}$$

$$L^\infty\text{-norm} : \|x\|_\infty = \max_{1 \le i \le n} (|x_i|)$$

Here the L^2-norm induces the standard Euclidean metric discussed above. The L^1-norm induces a metric known as the *Manhattan* or *Taxicab* metric, as it gives the distance travelled between two points in a city consisting of a grid of horizontal and vertical streets. The limit of the L^p-norms, the L^∞-norm, is simply equal to the modulus of the largest component. This is a norm we will use frequently in later chapters.

Inner product space

An inner product space is simply a vector space V endowed with an inner product. An inner product is a function $\langle \cdot, \cdot \rangle : V \times V \to \mathbb{R}$. As usual, the inner product on Euclidean space $\mathbb{R}^n$ is familiar, and is called the *dot product*, or *scalar product*. For two vectors $v = (v_1, \ldots, v_n)$ and $w = (w_1, \ldots, w_n)$ this is given by $\langle v, w \rangle = v \cdot w = v_1 w_1 + \cdots + v_n w_n$. On other vector spaces the inner product is a generalization of the Euclidean dot product. An inner product adds the concept of *angle* to the concept of *length* provided by the norm discussed above.

Topological space

Endowing a space with some topology formalizes the notions of continuity and connectedness.

Definition 3.2.4 *A topological space is a set X together with a set $\mathcal{T}$ containing subsets of X, satisfying:*

1. *The empty set $\emptyset$ and X itself belong to $\mathcal{T}$*
2. *The intersection of two sets in $\mathcal{T}$ is in $\mathcal{T}$*
3. *The union of any collection of sets in $\mathcal{T}$ is in $\mathcal{T}$*

The sets in $\mathcal{T}$ are *open sets*, which are the fundamental elements in a topological space. We give a definition for open sets in a metric space in Definition 3.2.6. The family of all open sets in a metric space forms a topology on that space, and so every metric space is automatically a topological space. In particular, since Euclidean space $\mathbb{R}^n$ is a metric space, it is also a topological space. (However, the reverse is not true, and there are topological spaces which are not metric spaces.)

Manifolds, smooth and Riemannian, and tangent spaces

The importance of Euclidean space can be seen in the definition of a *manifold*. This is a technical object to define formally, but we give the standard heuristic definition, that a *manifold* is a topological space that looks locally like Euclidean space $\mathbb{R}^n$. Of course, Euclidean space itself gives a straightforward example of a manifold. Another example is a surface like a sphere (such as the Earth) which looks like a flat plane to a small enough observer (producing the impression that the Earth is flat). The same could be said of other sufficiently well-behaved surfaces, such as the torus. We will not be concerned with spaces that are not manifolds.

The formal definition of a manifold involves local coordinate systems, or *charts*, to make precise the notion of 'looks locally like'. If these charts possess some regularity with respect to each other, we may have the notion of differentiability on the manifold. In particular, with sufficient regularity, a manifold is said to be a *smooth*, or *infinitely differentiable* manifold. Again, for all our purposes we will work with smooth manifolds.

On a smooth manifold M one can give the description of *tangent space*. Thus at each point $x \in M$ we associate a vector space (called tangent space, and written $T_x M$) which contains all directions in which a curve in M can pass through x. Elements in $T_x M$ are called *tangent vectors*, and these formalize the idea of directional derivatives. We will frequently have to work with the tangent space to describe the rate at which points on a manifold are separated. However, throughout the book our manifolds will be well-behaved two-dimensional surfaces, and so tangent space will simply be expressed in the usual Cartesian or polar coordinates.

Finally, if a differentiable manifold is such that all tangent spaces are equipped with an inner product then the manifold is said to be *Riemannian*. This allows a variety of notions, such as length, angle, volume, curvature, gradient and divergence. Again, all the domains we consider here will be Riemannian manifolds (or, as discussed in Chapter 5, a collection of Riemannian manifolds 'glued' together).

3.2.2 Describing sets of points

Once a metric is defined on a space (i.e., set of points), then it can be used to characterize other types of sets in the space, for example the sets of points 'close enough' to a given point:

Definition 3.2.5 (Open ϵ-ball) *The set*

$$B(x, \epsilon) = \{y \in M \,|\, d(x, y) < \epsilon\},$$

is called the open ϵ-ball around x.

Intuitively, such a set is regarded as *open*, as although it does not contain points *y* a distance of *exactly* ϵ away from *x*, we can always find another point in the set (slightly) further away than any point already in the set.

With this definition we can now define the notion of an open set.

Definition 3.2.6 (Open set) *A set $U \subset M$ is said to be open if for every $x \in U$ there exists an $\epsilon > 0$ such that $B(x, \epsilon) \subset U$.*

Thus, open sets have the property that all points in the set are surrounded by points that are *in the set*. The reader should check that this definition implies that a union of open sets is an open set. The family of open sets give the required topology for *M* to be a topological space. The notion of a *neighbourhood of a point* is similar to that of open set.

Definition 3.2.7 (Neighbourhood of a point) *If $x \in M$ and U is an open set containing x, then U is said to be a neighbourhood of x.*

Definition 3.2.8 (Limit point) *Let $V \subset M$, and consider a point $p \in V$. We say that p is a limit point of V if every neighbourhood of p contains a point $q \neq p$ such that $q \in V$.*

The notion of a boundary point of a set will also be useful.

Definition 3.2.9 (Boundary point of a set) *Let $V \subset M$. A point $x \subset V$ is said to be a boundary point of V if for every neighborhood U of x we have $U \cap V \neq \emptyset$ and $U \backslash V \neq \emptyset$ (where $U \backslash V$ means 'the set of points in U that are not in V ').*

So a boundary point of a set is *not* surrounded by points in the set, in the sense that you cannot find a neighbourhood of a boundary point a set having the property that the neighbourhood is in the set.

Definition 3.2.10 (Boundary of a set) *The set of boundary points of a set V is called the boundary of V, and is denoted ∂V.*

It is natural to define the interior of a set as the set that you obtain after removing the boundary. This is made precise in the following definition.

Definition 3.2.11 (Interior of a set) *For a set $V \subset M$, the interior of V, denoted Int V, is the union of all open sets that are contained in V. Equivalently, it is the set of all $x \subset V$ having the property that $B(x, \epsilon) \subset V$, for some $\epsilon > 0$. Equivalently, $\mathrm{Int} V = V \backslash \partial V$.*

Definition 3.2.12 (Closure of a set) *For a set $V \subset M$, the closure of V, denoted $\bar{V}$ is the set of $x \subset M$ such that $B(x, \epsilon) \cap V \neq \emptyset$ for all $\epsilon > 0$.*

So the closure of a set *V* may contain points that are not part of *V*. This leads us to the next definition.

Definition 3.2.13 (Closed set) *A set $V \subset M$ is said to be closed if $V = \bar{V}$.*

In the above definitions the notion of the complement of a set arose naturally. We give a formal definition of this notion.

Definition 3.2.14 (Complement of a set) *Consider a set $V \subset M$. The complement of V, denoted $M \backslash V$ (or V^c, or $M - V$) is the set of all points $p \in M$ such that $p \notin V$.*

Cantor sets play an important role in chaotic dynamical systems. We will see that the following definition is a key feature in the definition of a Cantor set.

Definition 3.2.15 (Perfect set) *A set $V \subset M$ is said to be perfect if it is closed, and if every point of V is a limit point of V.*

Given a 'blob' (i.e., set) in our flow domain (M), we will want to develop ways of quantifying how it 'fills out' the domain. We begin with some very primitive notions.

Definition 3.2.16 (Dense set) *A set $V \subset M$ is said to be dense in M if $\bar{V} = M$.*

Intuitively, while a dense set V may not contain *all* points of M, it does contain 'enough' points to be *close* to all points in M.

Definition 3.2.17 (Nowhere dense set) *A set $V \subset M$ is said to be nowhere dense if $\bar{V}$ has empty interior, i.e., it contains no (nonempty) open sets.*

The notion of *convergence of a sequence* should be (relatively) familiar.

Definition 3.2.18 (Convergent sequence) *A sequence $\{x_n\}_{n \in \mathbb{N}}$ in M is said to converge to $x \in M$ if for every $\epsilon > 0$ there exists an $N \in \mathbb{N}$ such that for every $n \geq N$ we have $d(x_n, x) < \epsilon$.*

3.2.3 Compactness and connectedness

In ergodic theory a property of metric spaces called *compactness* often plays an important role in certain constructions. We define this property.

Definition 3.2.19 (Compact metric space) *Suppose M is a metric space and let $\{U_i\}_{i \in I}$ denote a collection of open sets in M, where I is an index set, having the property that*

$$M \subset \bigcup_{i \in I} U_i.$$

If we can extract a finite number of open sets from this collection of open sets, $\{U_{i_1}, U_{i_2}, \ldots, U_{i_n}\}$ *such that*

$$M \subset \bigcup_{i_\ell \in I} U_{i_\ell},$$

then M is said to be compact. The collection $\{U_i\}_{i \in I}$ *is said to be an 'open cover' of M. So M is compact if every open cover has a finite subcover.*

From this definition it can be proven that compact metric spaces are closed and bounded. In fact, if the metric space is Euclidean (i.e., $\mathbb{R}^n$ with the Euclidean metric) then it is compact *if and only if* it is closed and bounded (this is the statement of the Heine–Borel theorem). This link between boundedness and compactness is central to the work in this book. The boundedness property of the flow domain is natural for us since we will be considering mixers with boundaries. Even if we were to consider an 'infinite' channel, our interests will be in the bounded cross-sectional flow. Also, all the theoretical results we discuss in later chapters are formulated for compact metric spaces, and in particular linked twist maps are defined on spaces which are naturally compact and metrizable (i.e., possessing a metric).

The notion of 'connectedness' of a region of a flow should be intuitively clear. Nevertheless, we provide a mathematical definition in the spirit of the above.

Definition 3.2.20 (Connected set) *A set* $V \subset M$ *is said to be connected if there do not exist two disjoint open subsets,* $A \subset M$, $B \subset M$ *such that A intersects V, B intersects V, and* $V \subset A \cup B$.

Definition 3.2.21 (Totally disconnected set) *A set* $V \subset M$ *is said to be totally disconnected if the only connected subsets are sets consisting of a single point.*

Key point: The dynamical systems we discuss lie on either the 2-torus, or a subset of the plane $\mathbb{R}^2$, both of which are naturally smooth compact connected Riemannian manifolds and so possess all the structure we will require for later mathematical results.

3.2.4 Measuring the 'size' of sets

We now return to the notion of quantifying specific blobs in the flow domain. The area or volume occupied by the blob is an important quantity. This notion fits into the much larger subject of *measure theory*. It is almost impossible to approach the ergodic theory literature without some understanding of the terminology and ideas of measure theory.

A *measure* is a function that assigns a (non-negative) number to a given set. The assigned number can be thought of as a size, probability or volume of the set. Indeed a measure is often regarded as a generalization of the idea of the volume (or area) of a set. Every definition of integration is based on a particular measure, and a measure could also be thought of as a type of 'weighting' for integration. To specify any function, including a measure, we must first specify the domain of the function. It is not the case that any arbitrary subset of M can be assigned a measure, and some restriction on the class of subsets for which the notion of volume makes sense is required.

The collection of subsets of M on which the measure is defined is called a σ-algebra over M. Briefly, a σ-algebra over M is a collection of subsets that is closed under the formation of countable unions of sets and the formation of complements of sets. More precisely, we have the following definition.

Definition 3.2.22 (σ-algebra over M) *A σ-algebra, $\mathcal{A}$, is a collection of subsets of M such that:*

1. $M \in \mathcal{A}$,
2. $M \backslash A \in \mathcal{A}$ *for* $A \in \mathcal{A}$,
3. $\bigcup_{n \geq 0} A_n \in \mathcal{A}$ *for all* $A_n \in \mathcal{A}$ *forming a finite or infinite sequence* $\{A_n\}$ *of subsets of M.*

In other words, a σ-algebra contains the space M itself, and sets created under countable set operations. We shall see shortly that a σ-algebra contains sets which can be *measured*. It is natural to ask how the open sets defined by the metric on M relate to the σ-algebra. It can be proved that for any collection of sets, there is a smallest σ-algebra containing that collection of sets (Rudin (1974)). The σ-algebra obtained in this way is said to be *generated* by this collection of sets. If we take as the collection of sets all open sets in M, then the smallest σ-algebra containing the open sets is called the *Borel σ-algebra*, $\mathcal{B}$. The sets in $\mathcal{B}$ are called *Borel sets*. From Definition 3.2.22, open sets, closed sets, countable unions of closed sets, and countable intersections of open sets are all Borel sets.

Now we can define the notion of a measure.

Definition 3.2.23 (Measure) *A measure μ is a real-valued function defined on a σ-algebra satisfying the following properties:*

1. $\mu(\emptyset) = 0$,
2. $\mu(A) \geq 0$,
3. *for a countable collection of disjoint sets* $\{A_n\}$, $\mu(\bigcup A_n) = \sum \mu(A_n)$.

These properties are easily understood in the context of the most familiar of measures. In two dimensions, area (or volume in three dimensions) intuitively has the following properties: the area of an empty set is zero; the area of any set is non-negative; the area of the union of disjoint sets is equal to the sum of the area of the constituent sets. The measure which formalizes the concept of area or volume in Euclidean space is called *Lebesgue measure*.

If μ is always finite, we can normalize to $\mu(M) = 1$. In this case there is an analogy with probability theory that is often useful to exploit, and μ is referred to as a *probability measure*. A set equipped with a σ-algebra is called a *measurable space*. If it is also equipped with a measure, then it is called a *measure space*.

There are a number of measure-theoretic concepts we will encounter in later chapters. The most common is perhaps the idea of a set of *zero measure*. We will repeatedly be required to prove that points in a set possess a certain property. In fact what we actually prove is not that *every* point in a set possesses that property, but that *almost every* point possesses the property. The exceptional points which fail to satisfy the property form a set of measure zero, and such a set is, in a measure-theoretic sense, negligible. Naturally, a subset $U \subset M$ has measure zero if $\mu(U) = 0$. Strictly speaking, we should state that U has measure zero *with respect to the measure* μ. Moreover, if $U \notin \mathcal{A}$, the σ-algebra, then it is not measurable and we must replace the definition by: a subset $U \subset M$ has μ-measure zero if there exists a set $A \in \mathcal{A}$ such that $U \subset A$ and $\mu(A) = 0$. However, in this book, and in the applications concerned, we will allow ourselves to talk about sets of measure zero, assuming that all such sets are measurable, and that the measure is understood. Note that a set of zero measure is frequently referred to as a *null set*. When referring to 'all points except for a set of zero measure', the phrase 'almost every point', or 'almost everywhere' is generally used, often abbreviated to 'a.e.'.

From the point of view of measure theory two sets are considered to be 'the same' if they 'differ by a set of measure zero'. This sounds straightforward, but to make this idea mathematically precise requires some effort. The mathematical notion that we need is the *symmetric difference* of two sets. This is the set of points that belong to exactly one of the two sets. Suppose $U_1, U_2 \subset M$; then the symmetric difference of U_1 and U_2, denoted $U_1 \triangle U_2$, is defined as $U_1 \triangle U_2 \equiv (U_1 \backslash U_2) \cup (U_2 \backslash U_1)$. We say that U_1 and U_2 are *equivalent (mod 0)* if their symmetric difference has measure zero.

This allows us to define precisely the notions of sets of *full measure* and *positive measure*. Suppose $U \subset M$, then U is said to have full measure if U and M are equivalent (mod 0). Intuitively, a set U has full measure in M if $\mu(U) = 1$

(assuming $\mu(M) = 1$). A set of positive measure is intuitively understood as a set $V \subset M$ such that $\mu(V) > 0$, that is, strictly greater than zero. The *support* of a measure μ on a metric space M is the set of all points $x \in M$ such that every open neighbourhood of x has positive measure.

Finally in this section we mention the notion of *absolute continuity* of a measure. If μ and v are two measures on the same measurable space M then v is absolutely continuous with respect to μ, written $v \ll \mu$, if $v(A) = 0$ for every measurable $A \subset M$ for which $\mu(A) = 0$. Although absolute continuity is not something which we will have to work with directly, it does form the basis of many arguments in ergodic theory, and in particular underpins the main theorems of Chapter 5. Its importance stems from the fact that for physical relevance we would like properties to hold on sets of positive Lebesgue measure, since Lebesgue measure corresponds to volume. Suppose however that we can only prove the existence of desirable properties for a different measure v. We would then like to show that v is absolutely continuous with respect to Lebesgue measure, as the definition of absolute continuity would guarantee that if $v(A) > 0$ for a measurable set A, then the Lebesgue measure of A would also be strictly positive. In other words, any property exhibited on a significant set with respect to v would also manifest itself on a significant set with respect to Lebesgue measure.

3.3 Mathematical ideas for describing the movement of blobs in the flow domain

In reading the dynamical systems or ergodic theory literature one encounters a plethora of terms describing transformations, e.g. isomorphisms, automorphisms, endomorphisms, homeomorphisms, diffeomorphisms, etc. In some cases, depending on the structure of the space on which the map is defined, some of these terms may be synonyms. Here we will provide a guide for this terminology, as well as describe what is essential for our specific needs.

First, we start very basically. Let A and B be arbitrary sets, and consider a map, mapping, function, or transformation (these terms are often used synonymously), $f : A \rightarrow B$. The key defining property of a function is that for each point $a \in A$, it has only one *image* under f, i.e., $f(a)$ is a *unique* point in B. Now f is said to be *one-to-one* if any two different points are *not* mapped to the same point, i.e., $a \neq a' \Rightarrow f(a) \neq f(a')$, and it is said to be *onto* if every point $b \in B$ has a *preimage* in A, i.e., for any $b \in B$ there is at least one $a \in A$ such that $f(a) = b$. These two properties of maps are important because necessary and sufficient conditions for a map to have an inverse[1] f^{-1} is that it be one-to-one and onto. There is synonomous terminology for these properties. A mapping that is one-to-one is said to be *injective* (and may be referred to as an *injection*), a mapping that is onto is said to be *surjective* (and may be referred to as a *surjection*), and a mapping that is both one-to-one and onto is said to be *bijective* (and may be referred to as a *bijection*).

So far we have talked about properties of the mapping alone, with no mention of the properties of the sets A and B. In applications, additional properties are essential for discussing basic properties such as continuity and differentiability. In turn, when we endow A and B with the types of structure discussed in the previous section, it then becomes natural to require the map to respect this structure, in a precise mathematical sense. In particular, if A and B are equipped with algebraic structures, then a bijective mapping from A to B that preserves the algebraic structures in A and B is referred to as an *isomorphism* (if $A = B$ then it is referred to as an *automorphism*). If A and B are equipped with a topological structure, then a bijective mapping that preserves the topological structure is referred to as a *homeomorphism*. Equivalently, a homeomorphism is a map f that is continuous and invertible with a continuous inverse. If A and B are equipped with a differentiable structure, then a bijective mapping that preserves the differentiable structure is referred to as a *diffeomorphism*. Equivalently, a diffeomorphism is a map that is differentiable and invertible with a differentiable inverse.

The notion of *measurability* of a map follows a similar line of reasoning. We equip A with a σ-algebra $\mathcal{A}$ and B with a σ-algebra $\mathcal{A}'$. Then a map $f : A \to B$ is said to be measurable (with respect to $\mathcal{A}$ and $\mathcal{A}'$) if $f^{-1}(A') \in \mathcal{A}$ for every $A' \in \mathcal{A}'$. In the framework of using ergodic theory to describe fluid mixing, it is natural to consider a measure space M with the Borel σ-algebra. It is shown in most analysis courses following the approach of measure theory that continuous functions are measurable. Hence, a diffeomorphism $f : M \to M$ is certainly also measurable. However, in considering properties of functions in the context of a measure space, it is usual to disregard, to an extent, sets of zero measure. To be more precise, many properties of interest (e.g. non-zero Lyapunov exponents, or f being at least two times continuously differentiable) may fail on certain exceptional sets. These exceptional sets will have zero measure, and so throughout this book transformations will be sufficiently well-behaved, including being measurable and sufficiently differentiable.

---

[1] The inverse of a function $f(x)$ is written $f^{-1}(x)$ and is defined by

$$f(f^{-1}(x)) = f^{-1}(f(x)) = x$$

Measure-preserving transformations

Next we can define the notion of a *measure-preserving transformation*.

Definition 3.3.1 (Measure-preserving transformation) *A transformation f is measure-preserving if for any measurable set $A \subset M$:*

$$\mu(f^{-1}(A)) = \mu(A) \text{ for all } A \in \mathcal{A}.$$

This is equivalent to calling the measure μ *f-invariant* (or simply *invariant*). If the transformation f is invertible (that is, f^{-1} exists), as in all the examples that we will consider, this definition can be replaced by the more intuitive definition.

Definition 3.3.2 *An invertible transformation f is measure-preserving if for any measurable set $A \subset M$:*

$$\mu(f(A)) = \mu(A) \text{ for all } A \in \mathcal{A}.$$

For those working in applications the notation $f^{-1}(A)$ may seem a bit strange when at the same time we state that it applies in the case when f is not invertible. However, it is important to understand $f^{-1}(A)$ from the point of view of its set-theoretic meaning: literally, it is the set of points that map to A under f. This does not require f to be invertible (and it could consist of disconnected pieces). We have said nothing so far about whether such an invariant measure μ might exist for a given transformation f, but a standard theorem, called the Kryloff–Bogoliouboff theorem (see for example Katok & Hasselblatt (1995), or Kryloff & Bogoliouboff (1937) for the original proof) guarantees that if f is continuous and M is a compact metric space then an invariant Borel probability measure does indeed exist.

In many of the examples that we will consider the measure of interest will be the area, i.e., the function that assigns the area to a chosen set. The fluid flow will preserve this measure as a consequence of the flow being incompressible. For two-dimensional blinking flows this is straightforward. For three-dimensional duct flows a little more thought may be required. For a three-dimensional incompressible flow it is perhaps not immediately obvious that the map of fluid particles between cross-sectional elements gives rise to an area-preserving map. However, for the duct flows that we consider this will be the case. Moreover, it will also be true for the linked twist maps that we consider, and one can verify directly that the LTMs we derived for duct flows in Chapter 2 are indeed area preserving (i.e., their determinant is identically one).

Finally, we end this section by pulling together the crucial concepts above into one definition.

Definition 3.3.3 (Measure-preserving dynamical system) *A measure-preserving dynamical system is a quadruple* $(M, \mathcal{A}, f, \mu)$ *consisting of a metric space M, a σ-algebra $\mathcal{A}$ over M, a transformation f of M into M, and a f-invariant measure μ.*

Key point: All of the transformations we will study will be sufficiently well-behaved. In particular, we will have all the invertibility, measurability and differentiability we require. Moreover, Lebesgue measure μ will be preserved, and we will always have $\mu(M) < \infty$, that is, M will be of finite measure.

3.4 Dynamical systems terminology and concepts

The following section reviews definitions that are needed in the rest of the book. What we will say is mathematically rigorous; however, we will strive to not make the presentation needlessly formal and occasionally we will offer physical interpretations of the definitions. The relevant background from dynamical systems theory can be found in Wiggins (2003), and many other places as well, while the kinematical aspects of mixing are covered in Ottino (1989a).

3.4.1 Terminology for general fluid kinematics

Mappings The motion of fluid particles is described mathematically with a map, mapping, or transformation. Let M denote the region occupied by the fluid. We refer to points in M as fluid particles. The flow of fluid particles is mathematically described by a smooth, invertible transformation, or map, of M into M, denoted by f, also having a smooth inverse. The particles are labelled by their initial condition at some arbitrary time, usually taken as $t = 0$. The application of f to the domain M, denoted $f(M)$, is referred to as one *advection cycle*. Similarly, n advection cycles are obtained by n repeated applications of f, denoted $f^n(M)$. Let A denote any subdomain of M. Then $\mu(A)$ denotes the volume of M (if we are in a 2D setting, read 'volume' as 'area'). Thus μ is a function that assigns to any (mathematically well-behaved) subdomain of M its volume. In the mathematical terminology developed above, the function μ is known as a measure, and the subdomain as measurable. Incompressibility of the fluid is expressed by stating that as any subdomain of M is stirred, its volume remains unchanged, i.e., $\mu(A) = \mu(f(A))$. In the language of dynamical systems theory, f is an example of a measure-preserving transformation. We will assume that M has finite volume, which is natural for the applications we have in mind. As described above, in this case we can normalize the function that assigns the volume to a subdomain of M, so that without loss of generality we can assume $\mu(M) = 1$. This will make certain mathematical definitions simpler, and is a standard assumption in the mathematics literature.

Orbits For a specific fluid particle p, the trajectory, or orbit, of p is the sequence of points $\{\ldots f^{-n}(p),\ldots,f^{-1}(p),p,f(p),f^2(p),\ldots,f^n(p),\ldots\}$. So the orbit of a point is simply the sequence of points corresponding to the point itself, where the point has been (under the past advection cycles), and where the point will go (under future advection cycles).

3.4.2 Specific types of orbits

The next four definitions refer to specific orbits that often have a special significance for transport and mixing.

Periodic orbit An orbit consisting of a finite number of points is called a periodic orbit (where the number of points is the period of the orbit). Such an orbit has the property that during each application of the advection cycle each point on the orbit shifts to another point on the orbit. Periodic orbits may be distinguished by their stability type. In the typical case, periodic orbits are either stable (i.e. nearby orbits remain near the periodic orbit), referred to as elliptic, or unstable of saddle type (i.e., meaning that typical nearby orbits either move away from the periodic orbit, or approach the periodic orbit for a time, but ultimately move away), referred to as hyperbolic. If one is interested in the design of a micromixer it may not seem particularly relevant to focus on particles that undergo periodic motion during the advection cycle. However, they are often the template of the global mixing properties. For example, elliptic periodic orbits are bad for mixing as they give rise to regions that do not mix with the surrounding fluid ('islands'). Hyperbolic periodic orbits provide mechanisms for contraction and expansion of fluid elements, and they can also play a central role in the existence of Smale horseshoe maps, which may lead to efficient global mixing properties.

Homoclinic orbit This is an orbit that, asymptotically in positive time (i.e., forward advection cycles), approaches a hyperbolic periodic orbit, and asymptotically in negative time (i.e. inverse advection cycles) approaches the same periodic orbit. These types of orbits are significant because in a neighbourhood of such an orbit a Smale horseshoe map can be constructed (the 'Smale–Birkhoff homoclinic' theorem).

Heteroclinic orbits and cycles This is an orbit that, asymptotically in positive time (i.e., forward advection cycles), approaches a hyperbolic periodic orbit, and asymptotically in negative time (i.e. inverse advection cycles) approaches a different periodic orbit. If two or more heteroclinic orbits exist and are arranged in a heteroclinic cycle, then it is generally possible to construct a Smale horseshoe map (this will be discussed in detail in Chapter 4) near the heteroclinic cycle in the same way as near a homoclinic orbit.

Chaotic trajectory There is still no universally accepted definition of the notion of chaos. Different approaches include the ideas of sensitive dependence on initial conditions; positive Lyapunov exponents; positive topological entropy; denseness of periodic orbits. We do not need to involve ourselves in the discussion of which definition is the most appropriate–for our purposes, chaotic behaviour is characterized by apparently unpredictable behaviour of increasing complexity in which typical initial conditions close to each other are separated at an exponential rate under the action of the dynamical system in question. Each definition of chaos in the literature expresses this type of behaviour to a greater or lesser extent. For us, the precision and rigour of definitions will appear in the ergodic theory of mixing, which goes hand in hand with the notion of chaos.

3.4.3 Behaviour near a specific orbit

Lyapunov exponents are a ubiquitous diagnostic in the chaotic dynamics literature. We will discuss them rigorously in Chapter 5, and here give a brief outline. It is important to understand that they are numbers associated with one orbit. Additional information (such as ergodicity, to be discussed below) may allow us to extend this knowledge to larger regions of the domain of the map.

Lyapunov exponents These are numbers associated with an orbit that describes its stability in the linear approximation (i.e., the growth rate of 'infinitesimal' perturbations). Elliptic periodic orbits have zero Lyapunov exponents. Hyperbolic periodic orbits have some positive and some negative Lyapunov exponents. In incompressible flows, the sum of all Lyapunov exponents for an orbit must be zero. It is important to realize that a Lyapunov exponent is an infinite time average. Consequently, it can only be approximated in general. Various people have considered so-called finite time Lyapunov exponents (see, e.g., Lapeyre (2002)), however, it is important to understand that they are not on the same rigorous mathematical footing as standard Lyapunov exponents (Oseledec (1968)), and their applicability to mixing must often be assessed on a case-by-case basis.

The limitations associated with the fact that Lyapunov exponents characterize infinitesimal separations of orbits have been addressed with the development of finite size Lyapunov exponents (see, e.g., Boffeta *et al.* (2001)). This is an interesting, and potentially important, development, but it should be realized that, at present, there are no rigourous mathematical foundations for this concept. It is mainly a computational tool whose effectiveness must be addressed on a case-by-case basis although it should be noted that work of Yomdin (1987) and Newhouse (1988) on the growth of curves in 2D flows, and curves and surfaces in 3D flows, is certainly relevant to our needs.

3.4.4 Sets of fluid particles that give rise to 'flow structures'

'Flow structures' are routinely and commonly seen in many flow visualization experiments. The essential goal of flow visualization experiments is to capture various types of flow structures. Dynamical systems theory provides a way of describing what is seen in these experiments, and also predicting their evolution, and dependence on parameters.

Invariant set Let A be a subdomain of M. Then A is said to be *invariant* under the advection cycle if $f(A) = A$. That is, all points in A remain in A under repeated applications of the advection cycle. Clearly, invariant sets strictly smaller than M are bad for mixing (see the definition of ergodicity that we will shortly give) since they represent subdomains of the flow that do not mix with the rest of M, except via the mechanism of molecular diffusion (but we are restricting our discussion solely to kinematical mechanisms for mixing). The orbit of a point p and a homoclinic orbit are examples of invariant sets (but ones with zero volume).

A dense orbit This idea may seem more appropriate in the category above labelled 'Specific types of orbits'. Indeed, it is a specific type of orbit, but its character derives from its relationship to the ambient space which contains it. In particular, to describe it we need the notion of an invariant set, which we will denote by A. Consider an orbit of f in A (since A is invariant the entire orbit must be in A; this is a key point). The orbit is referred to as a *dense orbit* if it forms a dense set (see Definition 3.2.16) in A. Sometimes one just reads, or hears, the phrase 'dense orbit'. By itself, this is not adequate for describing the situation. One must consider whether or not the orbit forms a dense set within a particular set.

KAM theorem The Kolmogorov–Arnold–Moser (KAM) theorem is concerned with the existence of quasiperiodic orbits in perturbations of integrable Hamiltonian systems, or volume preserving maps. These orbits have the property that they densely fill out tori or 'tubes'. These tubes are therefore material surfaces, and fluid particles cannot cross them. Consequently these tubes trap regions of fluid that cannot mix with their surroundings (without molecular diffusion). The theorem may seem essentially useless for direct application in the sense that it is rare to be in a situation of 'almost integrable'. Moreover, even if that were the case, rigorous verification of the hypotheses may be quite difficult, and even impossible. The theorem does imply the existence of islands in the neighbourhood of an elliptic periodic orbit (discussed below). In this sense, the KAM theorem is surprisingly effective and describes a phenomenon that has been observed to occur very generally in Hamiltonian systems and is present in virtually every computed example of Poincaré sections in area preserving maps. It has become traditional to refer to all such material tubes or tori as 'KAM tori', even if they are observed in situations where the theorem does not rigorously apply, or cannot be applied. Such tubes have been observed experimentally (Kusch & Ottino (1992), Fountain *et al.* (1998)).

Island Elliptic periodic orbits are significant because, according to the Kolmogorov–Arnold–Moser (KAM) theorem, they are surrounded by 'tubes' which trap fluid. Moreover, these tubes exhibit a strong effect on particles outside the tube, but close to the tube. In a mathematically rigorous sense, these tubes are 'sticky' (Perry & Wiggins (1994)). The tubes and the neighbouring region that they influence in this way are referred to as 'islands'. Clearly, islands inhibit good mixing.

Barriers to transport and mixing In certain circumstances there can exist surfaces of one less dimension than the domain M that are made up entirely of trajectories of fluid particles, i.e., material surfaces. Consequently, fluid particle trajectories cannot cross such surfaces and in this way they are barriers to transport. KAM tori are examples of complete barriers to transport: fluid particle trajectories starting inside remain inside forever. Partial barriers to transport are associated with hyperbolic periodic orbits. The collection of fluid particle trajectories that approach the periodic orbit asymptotically as time goes to positive infinity form a material surface called the stable manifold of the periodic orbit. Similarly, the collection of fluid particle trajectories that approach the periodic orbit asymptotically as time goes to negative infinity form a material surface called the unstable manifold of the periodic orbit. Homoclinic orbits can therefore be characterized as orbits that are in the intersection of the stable and unstable manifolds of a periodic orbit. Similarly, heteroclinic orbits can be characterized as orbits that are in the intersection of the stable manifold of one periodic orbit with the unstable manifold of another periodic orbit.

Lobe dynamics As mentioned above, the stable and unstable manifolds of hyperbolic period orbits are material curves and, therefore, fluid particle trajectories cannot cross them. However, they can deform in very complicated ways and result in intricate flow structures. The resulting flow structure is the spatial, geometrical template on which the transport and mixing takes place in time. Lobe dynamics provides a way of describing the geometrical structure and quantifying the resulting transport. See Rom-Kedar & Wiggins (1990) and Wiggins (1992) for the general theory, Camassa & Wiggins (1991) and Horner *et al.* (2002) for an application to a time-dependent 2D cellular flow and experiments and Beigie *et al.* (1994) for further applications and development.

3.5 Fundamental results for measure-preserving dynamical systems

In this section we give two classical, and extremely fundamental results for dyamical systems which preserve an invariant measure. The ideas are a foundation of much of the theory which follows in later chapters. We begin with a theorem about *recurrence*.

Theorem 3.5.1 (Poincaré recurrence theorem) *Let $(M, \mathcal{A}, f, \mu)$ be a measure-preserving dynamical system (such that $\mu(M) < \infty$), and let $A \in \mathcal{A}$ be an arbitrary measurable set with $\mu(A) > 0$. Then for almost every $x \in A$, there exists $n \in \mathbb{N}$ such that $f^n(x) \in A$, and moreover, there exists infinitely many $k \in \mathbb{N}$ such that $f^k(x) \in A$.*

Proof Let B be the set of points in A which never return to A,

$$B = \{x \in A | f^n(x) \notin A \text{ for all } n > 0\}.$$

We could also write

$$B = A \setminus \cup_{i=0}^{\infty} f^{-n}(A).$$

First note that since $B \subseteq A$, if $x \in B$, then $f^n(x) \notin B$, by the definition of B. Hence $B \cap f^{-n}(B) = \emptyset$ for all $n > 0$ (if not then applying f^n contradicts the previous sentence). We also have $f^{-n}(B) \cap f^{n+k}(B) = \emptyset$ for all $n > 0, k \geq 0$ (else a point in $f^{-k}(B) \cap f^{-(n+k)}(B)$ would have to map under f^{-k} into both B and $f^{-n}(B)$ and we have just seen that these are disjoint). Therefore the sets $B, f^{-1}(B), f^{-2}(B), \ldots$ are pairwise disjoint. Moreover because f is measure-preserving $\mu(B) = \mu(f^{-1}(B)) = \mu(f^{-2}(B)) = \cdots$. Now we have a collection of an infinite number of pairwise disjoint sets of equal measure in M, and since $\mu(M) < 1$ we must have $\mu(B) = 0$, and so for almost every $x \in A$ we have $f^n(x) \in A$ for some $n > 0$. To prove that the orbit of x returns to A infinitely many times, we note that we can simply repeat the above argument starting at the point $f^n(x) \in A$ to find $n' > n$ such that $f^{n'}(x) \in A$ for almost every $x \in A$, and continue in this fashion. $\square$

One of the most important results concerning measure-preserving dynamical systems is the *Birkhoff Ergodic Theorem*, which tells us that for typical initial conditions, we can compute time averages of functions along an orbit. Such functions ϕ on an orbit are known as *observables*, and in practice might typically be a physical quantity to be measured, such as concentration of a fluid. On the theoretical side, it is crucial to specify the class of functions to which ϕ belongs. For example we might insist that ϕ be measurable, integrable, continuous or differentiable.

We give the theorem without proof, but the reader could consult, for example, Birkhoff (1931), Katznelson & Weiss (1982), Katok & Hasselblatt (1995) or Pollicott & Yuri (1998) for further discussion and proofs.

Theorem 3.5.2 (Birkhoff Ergodic Theorem) *Let* $(M, \mathcal{A}, f, \mu)$ *be a measure-preserving dynamical system, and let* $\phi \in \mathcal{L}^1$ *(i.e., the set of functions on M such that $\int_M |\phi| \mathrm{d}\mu$ is bounded) be an observable function. Then the forward time average* $\phi^+(x)$ *given by*

$$\phi^+(x) = \lim_{n\to\infty} \frac{1}{n} \sum_{i=0}^{n-1} \phi(f^i(x)) \qquad (3.1)$$

exists for μ-almost every $x \in M$. Moreover, if $\mu(M) < \infty$, the time average ϕ^+ *satisfies*

$$\int_M \phi^+(x)\mathrm{d}\mu = \int_M \phi(x)\mathrm{d}\mu.$$

This theorem can be restated for negative time to show that the backward time average

$$\phi^-(x) = \lim_{n\to\infty} \frac{1}{n} \sum_{i=0}^{n-1} \phi(f^{-i}(x))$$

also exists for μ-almost every $x \in M$, and $\int_M \phi^-(x)\mathrm{d}\mu = \int_M \phi(x)\mathrm{d}\mu$ if $\mu(M) < \infty$. A simple argument reveals that forward time averages equal backward time averages almost everywhere.

Lemma 3.5.1 *Let* $(M\mathcal{A}, f, \mu)$ *be a measure-preserving dynamical system with* $\mu(M) < \infty$ *and let $\phi \in \mathcal{L}^1$ be an observable function. Then*

$$\phi^+(x) = \phi^-(x)$$

for almost every $x \in M$; that is, the functions ϕ^+ and ϕ^- coincide almost everywhere.

Proof Let $A^+ = \{x \in M | \phi^+(x) > \phi^-(x)\}$. By definition A^+ is an invariant set, since $\phi^+(x) = \phi^+(f(x))$. Thus applying the Birkhoff Ergodic Theorem to the transformation f restricted to the set A^+ we have

$$\int_{A^+} (\phi^+(x) - \phi^-(x))d\mu = \int_{A^+} \phi^+(x)d\mu - \int_{A^+} \phi^-(x)d\mu$$

$$= \int_{A^+} \phi(x)d\mu - \int_{A^+} \phi(x)d\mu$$

$$= 0.$$

Then since the integrand in the first integral is strictly positive by definition of A^+ we must have $\mu(A^+) = 0$, and so $\phi^+(x) \le \phi^-(x)$ for almost every $x \in M$. Similarly, the same argument applied to the set $A^- = \{x \in M | \phi^-(x) > \phi^+(x)\}$ implies that $\phi^-(x) \le \phi^+(x)$ for almost every $x \in M$, and so we conclude that $\phi^+(x) = \phi^-(x)$ for almost every $x \in M$. $\qquad\square$

The Birkhoff Ergodic Theorem tells us that forward time averages and backward time averages exist, providing we have an invariant measure. It also says that the spatial average of a time average of an integrable function ϕ is equal to the spatial average of ϕ. Note that it does *not* say that the time average of ϕ is equal to the spatial average of ϕ. For this to be the case, we require ergodicity.

3.6 Ergodicity

In this section we describe the notion of *ergodicity*, but first we emphasize an important point. Recall that we are assuming M is a compact metric space and that the measure of M is finite. Therefore all quantities of interest can be rescaled by $\mu(M)$. In this way, without loss of generality, we can take $\mu(M) = 1$. In the ergodic theory literature, this is usually stated from the start, and all definitions are given with this assumption. In order to make contact with this literature, we will follow this convention. However, in order to get meaningful estimates in applications one usually needs to take into account the size of the domain (i.e. $\mu(M)$). We will address this point when it is necessary.

There are many equivalent definitions of ergodicity. The basic idea is one of *indecomposability*. Suppose a transformation f on a space M was such that two sets of positive measure, A and B, were invariant under f. Then we would be justified in studying f restricted to A and B separately, as the invariance of A and B would guarantee that no interaction between the two sets occurred. For an ergodic transformation this cannot happen – that is, M cannot be broken down into two (or more) sets of positive measure on which the transformation may be studied separately. This need for the lack of non-trivial invariant sets motivates the definition of ergodicity.

Definition 3.6.1 (Ergodicity) *A measure-preserving (invertible) dynamical system* $(M, \mathcal{A}, f, \mu)$ *is ergodic if* $\mu(A) = 0$ *or* $\mu(A) = 1$ *for all* $A \in \mathcal{A}$ *such that* $f(A) = A$.

We sometimes say that f is an ergodic transformation, or that μ is an ergodic invariant measure.

Ergodicity is a measure-theoretic concept, and is sometimes referred to as *metrical transitivity*. This evokes the related idea from topological dynamics[2] of *topological transitivity* (which is sometimes referred to as *topological ergodicity*).

Definition 3.6.2 (Topological transitivity) *A (topological) dynamical system* $f : X \to X$ *is topologically transitive if for every pair of open sets* $U, V \subset X$ *there exists an integer n such that* $f^n(U) \cap V \neq \emptyset$.

A topologically transitive dynamical system is often defined as a system such that the forward orbit of some point is dense in X. These two definitions are in fact equivalent (for homeomorphisms on a compact metric space), a result given by the *Birkhoff Transitivity Theorem* (see for example, Robinson (1998)).

Another common, heuristic way to think of ergodicity is that orbits of typical initial conditions come arbitrarily close to every point in M, i.e. typical orbits are dense in M. 'Typical' means that the only trajectories not behaving in this way form a set of measure zero. More mathematically, this means the only invariant sets are trivial ones, consisting of sets of either full or zero measure. However, as we shall see in Section 3.6.1, ergodicity is a stronger property than the existence of dense orbits.

The importance of the concept of ergodicity to fluid mixing is clear. An invariant set by definition will not 'mix' with any other points in the domain, so it is vital that the only invariant sets either consist of negligibly small amounts of points, or comprise the whole domain itself (except for negligibly small amounts of points). The standard example of ergodicity in dynamical systems is the map consisting of rigid rotations of the circle.

Example 3.6.1 *Let* $M = \mathcal{S}^1$, *and* $f(x) = x + \omega$ (mod 1). *Then if*

$$\omega \text{ is a rational number, } f \text{ is not ergodic,}$$

$$\omega \text{ is an irrational number, } f \text{ is ergodic.}$$

A rigorous proof can be found in any book on ergodic theory, for example Petersen (1983). We note here that the irrational rotation on a circle is an

---

[2] For topological concepts we refer to a topological space X (recall Definition 3.2.4), while for measure-theoretic results we refer to a metric space M (recall Definition 3.2.2).

example of an even more special type of system. The infinite non-repeating decimal part of ω means that x can never return to its initial value, and so no periodic orbits are possible.

Example 3.6.2 (Lasota & Mackey (1994)) *Let* $U = [0, 1] \times [0, 1]$ *be the unit square and consider the measure-preserving transformation* $f : U \to U$ *given by*

$$f(x, y) = (x + \sqrt{2}, y + \sqrt{3}) \quad (\text{mod } 1),$$

which is known to be ergodic, but not mixing. The behaviour of this map is illustrated in Figure 3.1. The first figure (a) shows the result of iterating a single initial condition 2000 times. The ergodicity of the map can be seen by the fact that the trajectory visits the entire domain. The incommensurate nature of the map ensures that every trajectory is dense in the unit square. (The fact that every orbit is dense means there are no periodic orbits, and so only one invariant measure is supported.) However, there is no expansion or contraction in this map (it has zero Lyapunov exponents in the language of later chapters). This means that it has no mixing effect at all, despite the fact that individual trajectories traverse the whole domain. Figures 3.1(b)–(f) show the result of iterating 1000 points initially in $[0, 0.1] \times [0, 0.1]$ *(as in (b) – the top 500 coloured blue and the bottom 500 red) for 1, 2, 3 and 4 iterates (c,d,e and f respectively). While the 'blob' moves around the unit square, no mixing of red and blue takes place.*

There are a number of common ways to reformulate the definition of ergodicity. Indeed, these are often quoted as definitions. We give three of the most common here, as they express notions of ergodicity which will be useful in later chapters. The first two are based on the behaviour of observable functions for an ergodic system.

Definition 3.6.3 (Ergodicity – equivalent) *A measure-preserving dynamical system* $(M, \mathcal{A}, f, \mu)$ *is ergodic if and only if every invariant measurable (observable) function* ϕ *on M is constant almost everywhere.*

This is simply a reformulation of the definition in functional language, and it is not hard to see that this is equivalent to the earlier definition (see for example Katok & Hasselblatt (1995) or Brin & Stuck (2002) for a proof). We will make use of this equivalent definition later. Perhaps a more physically oriented notion of ergodicity comes from Boltzmann's development of statistical mechanics and is succinctly stated as 'time averages of observables equal space averages'. In other words, the long-term time average of a function ('observable') along a single 'typical' trajectory should equal the average of that function over all possible initial conditions. We state this more precisely below.

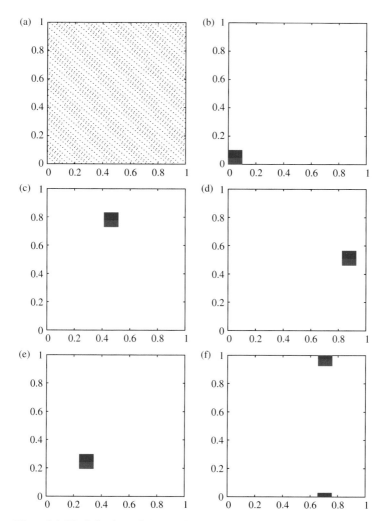

Figure 3.1 The behaviour of the ergodic map in Example 3.6.2. Figure (a) shows 200 iterates of a single initial condition. The trajectory is dense in the unit square. Figure (b) shows 1000 initial conditions coloured red and blue in $[0, 0.1] \times [0, 0.1]$. Figures (c)–(f) show the first four iterates of these initial conditions. No mixing occurs, but the 'blob' moves (also on a dense orbit) around the domain.

Definition 3.6.4 (Ergodicity – equivalent) *A measure-preserving dynamical system* $(M, \mathcal{A}, f, \mu)$ *is ergodic if and only if for all* $\phi \in \mathcal{L}^1$ *(i.e., the set of functions on M such that* $\int_M |\phi| \mathrm{d}\mu$ *is bounded), we have*

$$\lim_{n \to \infty} \frac{1}{n} \sum_{k=0}^{n-1} \phi(f^k(x)) = \int_M \phi \mathrm{d}\mu.$$

This definition deserves a few moments thought. The right-hand side is clearly just a constant, the spatial average of the function ϕ. It might appear that the left-hand side, the time average of ϕ along a trajectory depends on the given trajectory. But that would be inconsistent with the right-hand side of the equation being a constant. Therefore the time averages of typical trajectories are all equal, and are equal to the spatial average, for an ergodic system.

We have yet another definition of ergodicity that will 'look' very much like the definitions of mixing that we introduce in Section 3.7.

Definition 3.6.5 (Ergodicity – equivalent) *The measure-preserving dynamical system $(M, \mathcal{A}, f, \mu)$ is ergodic if and only if for all A, $B \in \mathcal{A}$,*

$$\lim_{n \to \infty} \frac{1}{n} \sum_{k=0}^{n-1} \mu(f^k(A) \cap B) = \mu(A)\mu(B).$$

It can be shown (see, e.g., Petersen (1983)) that each of these definitions imply each other (indeed, there are even more equivalent definitions that can be given). A variety of equivalent definitions is useful because verifying ergodicity for a specific dynamical system is notoriously difficult, and the form of certain definitions may make them easier to apply for certain dynamical systems.

3.6.1 A typical scheme for proving ergodicity

Ergodicity and topological transitivity are closely connected, and indeed for continuous maps, ergodicity implies topological transitivity.

Theorem 3.6.1 *Suppose a measure-preserving dynamical system $(M, \mathcal{A}, f, \mu)$ is ergodic. Then f is topologically transitive.*[3]

Proof Let $A \in \mathcal{A}$ be an open measurable set, so that $\mu(A) > 0$. The union $B = \cup_{k \in \mathbb{N}} f^{-k}(A)$ is by definition an invariant set, and so by the ergodicity of f has measure 0 or 1. Since $\mu(A) > 0$ and f is measure-preserving $\mu(B) = 1$. Therefore for almost every $x \in M$, $f^n(x) \in A$ for some $n \geq 0$. Since A is an arbitrary open set, almost every $x \in M$ visits any open set, and so we have a dense orbit, and hence topological transitivity. □

The converse of this theorem is not true – topological transitivity is a necessary but not sufficient condition for ergodicity.[4] One approach to proving

---

[3] In fact this proof gives a stronger result, that ergodicity implies that almost every orbit is dense, a property called *minimality*.

[4] An example of a map which is topologically transitive but not ergodic is the complex exponential map.

ergodicity is based on Definition 3.6.3, to show that all invariant measurable functions are constant almost everywhere. We first give the following definition.

Definition 3.6.6 (Locally constant almost everywhere) *An integrable function ϕ is locally constant almost everywhere if there exists a null set N (recall, a null set is a set of measure zero) such that for each $x \in M \backslash N$, ϕ is constant on almost every point in some neighbourhood of x.*

With this definition we can prove the following.

Theorem 3.6.2 *The measure-preserving transformation f is ergodic if*

1. *f is topologically transitive, and*
2. *for each continuous function ϕ, the forward time average $\phi^+(x)$ is locally constant almost everywhere.*

Proof Since $\phi^+(x)$ is locally constant almost everywhere, there exists a null set N such that for any pair of points $x, y \in M \backslash N$ we can find neighbourhoods $V(x)$ and $V(y)$ of x and y respectively such that $\phi^+(z)$ is constant (say $\phi^+(z) = C$) for almost every $z \in V(x)$, and $\phi^+(z')$ is constant for almost every $z' \in V(y)$. Without topological transitivity we may have $\phi^+(z) \neq \phi^+(z')$, but if f is topologically transitive, then by definition $f^n(V(x)) \cap V(y) \neq \emptyset$ for some integer n. Then,

$$
\begin{aligned}
C = \phi^+(z) \quad &\text{a.e.} \quad z \in V(x) \\
= \phi^+(z) \quad &\text{a.e.} \quad z \in f^n(V(x)) \\
= \phi^+(z) \quad &\text{a.e.} \quad z \in f^n(V(x)) \cap V(y) \\
= \phi^+(z') \quad &\text{a.e.} \quad z' \in V(y)
\end{aligned}
$$

Since this holds for almost every pair of points x and y, and for any locally constant almost everywhere continuous function ϕ, this shows ergodicity by Definition 3.6.3. See Figure 3.2. □

Theorem 3.6.2 reduces the task of proving ergodicity to showing the two conditions in the theorem. As we shall see in Chapter 5, linked twist maps possess some extra structure (a *hyperbolic* structure) which helps to show both topological transitivity and that time averages are almost everywhere locally constant.

Key point: Ergodicity is a desirable property from the point of view of fluid mixing as the indecomposable nature of an ergodic transformation means that islands cannot be formed. However, ergodic is not synonymous with chaotic, and ergodicity does not imply that regions of fluid will mix with each other.

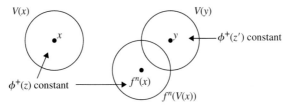

Figure 3.2 An illustration of how locally constant time averages combine with topological transitivity to produce ergodicity. The time average ϕ^+ is constant in a neighbourhood of x, and since time averages are invariant under the transformation, ϕ^+ equals the same constant in the neighbourhood $f^n(V(x))$ for each n. But by topological transitivity, for some n, $f^n(V(x))$ intersects $V(y)$, and so ϕ^+ equals the same constant in a neighbourhood of y.

3.7 Mixing

We now discuss ergodic theory notions of mixing, and contrast them with ergodicity. In the ergodic theory literature the term *mixing* is encompassed in a wide variety of definitions that describe different strengths or degrees of mixing. Frequently the difference between these definitions is only tangible in a theoretical framework. We give the most important definitions for applications (thus far) below.

Definition 3.7.1 ((Strong) Mixing) *A measure-preserving (invertible) transformation $f : M \to M$ is (strong) mixing if for any two measurable sets $A, B \subset M$ we have:*

$$\lim_{n \to \infty} \mu(f^n(A) \cap B) = \mu(A)\mu(B)$$

Again, for a non-invertible transformation we replace f^n in the definition with f^{-n}. The word 'strong' is frequently omitted from this definition, and we will follow this convention and refer to 'strong mixing' simply as 'mixing'. This is the most common, and the most intuitive definition of mixing, and we will describe the intuition behind the definition. To do this, we will *not* assume that $\mu(M) = 1$.

Within the domain M let A denote a region of, say, black fluid and let B denote any other region within M (see Figure 3.3). Mathematically, we denote the amount of black fluid that is contained in B after n applications of f by

$$\mu(f^n(A) \cap B),$$

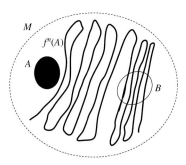

Figure 3.3 A sketch illustrating the principle of mixing. Under iteration of f, the set A spreads out over M, until the proportion of A found in any set B is the same as the proportion of A in M.

that is, the volume of $f^n(A)$ that ends up in B. Then the fraction of black fluid contained in B is given by

$$\frac{\mu(f^n(A) \cap B)}{\mu(B)}.$$

Intuitively, the definition of mixing should be that, as the number of applications of f is increased, for any region B we would have the same proportion of black fluid in B as the proportion of black fluid in M. That is,

$$\frac{\mu(f^n(A) \cap B)}{\mu(B)} - \frac{\mu(A)}{\mu(M)} \to 0, \text{ as } n \to \infty,$$

Now if we take $\mu(M) = 1$, we have

$$\mu(f^n(A) \cap B) - \mu(A)\mu(B) \to 0 \quad \text{as } n \to \infty,$$

which is our definition of (strong) mixing. Thinking of this in a probabilistic manner, this means that given any subdomain, upon iteration it becomes (asymptotically) independent of any other subdomain.

Like ergodicity, measure-theoretic mixing has a counterpart in topological dynamics, called *topological mixing*.

Definition 3.7.2 (Topological mixing) *A (topological) dynamical system $f :$ $X \to X$ is topologically mixing if for every pair of open sets $U, V \subset X$ there exists an integer $N > 0$ such that $f^n(U) \cap V \neq \emptyset$ for all $n \geq N$.*

Note the relationship between this definition and Definition 3.6.2. For topological transitivity we simply require that for any two open sets, an integer n (which will depend on the open sets in question) can be found such that the nth iterate of one set intersects the other. For topological mixing we require an integer N which is valid for all open sets, such that whenever $n \geq N$, the

nth iterate of one set intersects the other. Again the measure-theoretic concept of mixing is stronger than topological mixing, so that the following theorem holds, but not the converse.

Theorem 3.7.1 *Suppose a measure-preserving dynamical system* $(M, \mathcal{A}, f, \mu)$ *is mixing. Then f is topologically mixing.*

Proof See, for example, Petersen (1983).   □

Key point: Mixing is the central theme of this book and its ergodic theory definition is intuitively identical to the concept of mixing for fluids.

Definition 3.7.3 (Weak mixing) *The measure-preserving transformation f :* $M \to M$ *is said to be weak mixing if for any two measurable sets $A, B \subset M$ we have:*

$$\lim_{n \to \infty} \frac{1}{n} \sum_{k=0}^{n-1} |\mu(f^k(A) \cap B) - \mu(A)\mu(B)| = 0.$$

It is easy to see that weak mixing implies ergodicity (take A such that $f(A) = A$, and take $B = A$. Then $\mu(A) - \mu(A)^2 = 0$ implies $\mu(A) = 0$ or 1). The converse is not true, for example an irrational rotation is not weak mixing.

Mixing is a stronger notion than weak mixing, and indeed mixing implies weak mixing. Although the converse is not true it is difficult to find an example of a weak mixing system which is not mixing (such an example is constructed in Parry (1981)). The difference between weak and strong mixing appears unlikely to be distinguishable in applications.

It may seem that the notions of ergodicity and mixing are not suited for application to fluid mixing, since they are characterized by definitions involving infinite time limits, and we would like fluids to mix in something rather less than infinite time! However, these are useful concepts since (as discussed above) the existence of invariant sets and, in particular, 'islands', are typical mechanisms by which ergodicity and mixing break down, and these inhibit good fluid mixing. Moreover, in many areas of applications the *decay of correlations of a scalar field* is used as a diagnostic for quantifying mixing. We now show how this concept is related to the definition of mixing given above. First, we need a few technical definitions and notation. For a region B in M the function χ_B, referred to as the characteristic function associated with B, assigns 1 to a point that is in B, and 0 to all other points. Then the volume of B can be written as the integral $\int \chi_B d\mu = \mu(B)$ (think of $d\mu$ as the 'infinitesimal volume element' of integration, and then integrating over the region B). Similarly $\chi_{f^n(A) \cap B}$ assigns

1 to a point that is at the intersection of the nth mapping of A with B, and 0 to all other points. So in terms of integrals over characteristic functions, the limit in the definition of mixing above can be written as

$$\int \chi_{f^n(A)\cap B}\mathrm{d}\mu - \int \chi_A\mathrm{d}\mu \int \chi_B\mathrm{d}\mu \to 0 \quad \text{as} \quad n \to \infty,$$

and we can view this as an integral formulation of the mixing condition. This expression can be written in a more useful form. First, note that:

$$\chi_{f^n(A)\cap B} = \chi_B \chi_{f^n(A)}.$$

Now $f^n(A)$ is the set of points that map to A under f^{-n}. Therefore

$$\chi_{f^n(A)} = \chi_A \circ f^{-n}.$$

Putting all of this together, the integral expression derived above can be rewritten as

$$\int \chi_B(\chi_A \circ f^{-n})\mathrm{d}\mu - \int \chi_A\mathrm{d}\mu \int \chi_B\mathrm{d}\mu \to 0 \quad \text{as} \quad n \to \infty.$$

This last expression suggests a modification of the definition where we might replace the characteristic functions with arbitrary functions (from a certain class of functions of interest). So for functions ϕ and ψ (from the class of interest) we define the correlation function:

$$C_n(\phi, \psi) = \left| \int \phi(\psi \circ f^{-n})\mathrm{d}\mu - \int \phi\mathrm{d}\mu \int \psi\mathrm{d}\mu \right|.$$

In the language of ergodic theory, ϕ and ψ are referred to as observables, and the decay of the correlation function for general observables is considered. In applications, it is usually a specific observable that is of interest – one typically takes $\phi = \psi =$ 'a scalar field' (for example, concentration of a fluid), and considers the decay of correlations of that scalar field. Of course, if the transformation is not mixing then we should not expect the correlations to decay to zero. If the transformation is mixing, however, then the rate of decay of correlations is a quantifier of the speed of mixing. A study of the decay of correlations of different classes of mappings is currently at the forefront of research in dynamical systems theory. An excellent review is given by Baladi (2001).

Example 3.7.1 (Lasota & Mackey (1994)) *Consider the transformation g :
$U \to U$ given by*

$$g(x, y) = (x + y, x + 2y) \quad (\mathrm{mod}\ 1),$$

which is known to be mixing (and therefore ergodic). (This map is known as the Arnold Cat Map and will be revisited in greater detail in Chapter 5.) Like the *previous example it is an area-preserving map (the determinant of the Jacobian is equal to one), and we illustrate its behaviour in a similar way. Figure 3.4(a) shows a single trajectory of 2000 points. Again the orbit is dense on the unit square. Observe however that the trajectory fills the square in a less ordered way than for the previous example. Periodic orbits are possible in this map (the origin is clearly an example of a period-1 point). Figures 3.4(b) shows the 'blob' of 1000 points in $[0, 0.1] \times [0, 0.1]$ as before, with its image under 1,2,3 and 5 iterates in figures (c), (d), (e) and (f). Here the red and blue regions are quickly mixed up.*

Another simple example of a (strong) mixing system is the Baker's map. This is a two-dimensional map defined on the unit square $U = [0, 1] \times [0, 1]$. It is a useful example as it is intuitive, and is immediately amenable to analysis.

Example 3.7.2 *Example of mixing–the Baker's map*

Let $B : U \to U$ be a map of the unit square given by

$$B(x, y) = \begin{cases} (2x, y/2) & 0 \le x < 1/2 \\ (2x - 1, (y + 1)/2) & 1/2 \le x \le 1 \end{cases}$$

Note that this definition guarantees that both x and y remain in the interval $[0, 1]$. This map can be found defined in slightly different ways, using (mod 1) to guarantee this. The action of this map is illustrated in Figure 3.5. The name of the transformation comes from the fact that its action can be likened to the process of kneading bread when baking. At each iterate, the x variable is stretched by a factor of two, while the y variable is contracted by a factor of one half. This ensures that the map is area-preserving. Since the stretching in the x direction leaves the unit square, the right-hand half of the image cut off and placed on top of the left half. (This is similar to the Smale horseshoe discussed in the following chapter, except that there the map is kept in the unit square by folding (with a loss of material) rather than by cutting.)

Theorem 3.7.2 *The Baker's map of Example 3.7.2 is mixing.*

Proof This is simple to see diagrammatically. Figure 3.6 shows two sets X and Y in the unit square U. For simplicity we have chosen X to be the bottom half of the U, so that $\mu(X) = 1/2$, while Y is some arbitrary rectangle. The following five diagrams in Figure 3.6 show the image of X under 5 iterations of the Baker's map. It is clear that after n iterates X consists of 2^n strips of width $1/2^{n+1}$. It is then easy to see that

$$\lim_{n \to \infty} \mu(f^n(X) \cap Y) = \frac{\mu(Y)}{2} = \mu(X)\mu(Y).$$

A similar argument suffices for any two sets X and Y, and so the map is mixing. $\square$

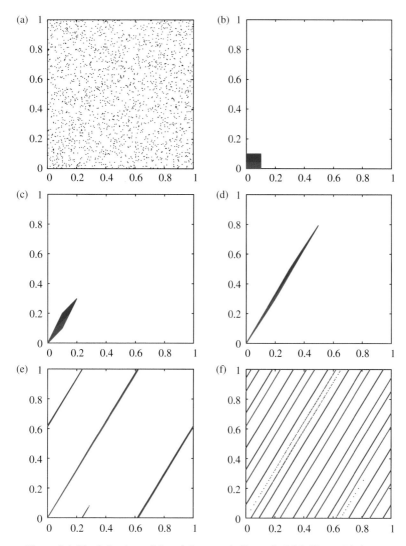

Figure 3.4 The behaviour of the mixing map in Example 3.7.1. Figure (a) shows 2000 iterates of a single initial condition. This trajectory is a dense orbit which fills the domain in an apparently disordered way. Figure (b) is a region of 1000 initial conditions coloured red and blue. Figures (c)–(f) show the image of the 'blob' after 1, 2, 3 and 5 iterations respectively. The red and blue regions are quickly mixed.

In general we cannot use this direct approach to prove mixing – in real systems mixing typically goes hand in hand with a huge increase in geometrical complexity and more subtle techniques are required.

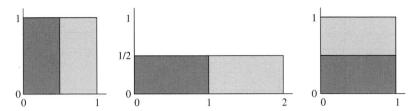

Figure 3.5 The action of the Baker's map of Example 3.7.2. The unit square U is stretched by a factor of two in the x direction while contracted by a factor of one half in the y direction. To return the image to U the right-hand half is cut off and placed on the left-hand half.

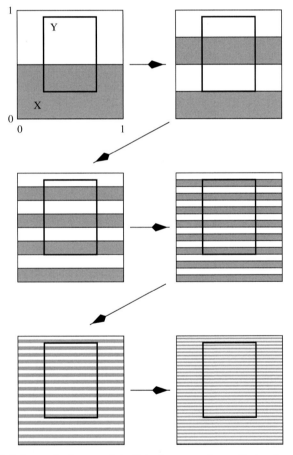

Figure 3.6 Successive iterates of the Baker's map on the set X, showing that the map is mixing.

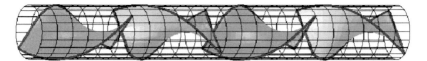

Figure 3.7 The Kenics®Mixer of Example 3.7.3 (picture taken from Galaktionov *et al.* (2003)). Blades positioned along a cylindrical pipe mimic the action of the Baker's map.

Example 3.7.3 *The Kenics®Mixer*

The Kenics®Mixer is a widely used static mixer based on the stretching, cutting and stacking action of the Baker's map. It consists of a cylindrical pipe containing mixing elements in the form of twisted blades. A typical configuration of the mixing elements is shown in Figure 3.7. Fluid is driven through the pipe, and is effectively stretched whilst passing a blade, and then cut and stacked at the interface between blades. Figure 3.8 illustrates this action in the cross-section of the pipe (as with Figure 3.7, this is taken from Galaktionov et al. (2003)). Figure (a) shows an initial concentration of fluid. The top half of the pipe is filled with white fluid, and the bottom half with black fluid. The vertical line represents the leading edge of the first blade. Figures (b), (c), (d) and (e) show the effect of the fluid passing the first blade. With the twist of the blade, the fluid is stretched until the original two stripes of black and white become four stripes. In (e) the position of the leading edge of the second blade is given as a dotted line. Figure (f) shows the start of the effect from the second blade. The material is cut and again stretched until the number of black and white stripes is again doubled, as in (g). This process continues, doubling the number of material striations with each mixing element, as in (h).

We note that the definition of mixing involves choosing two sets A and B, and then iterating one while keeping the other fixed. Mixing occurs if the iterated set 'spreads itself' over the domain. In the above examples, and in all subsequent illustrations of mixing or lack thereof, we choose two sets and iterate them both. The reason for this is that the 'spreading' of iterated sets can be as easily seen, but it is also more intuitive from the point of view of applications to see two blobs mix with each other.

3.8 The *K*-property

In reading the literature on ergodic theory and mixing it is almost certain that one will encounter the notion of the 'K-property'. The K-property fits between

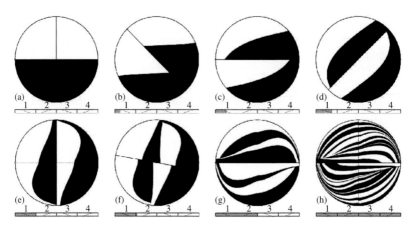

Figure 3.8 Concentration profiles for the Kenics®Mixer of Example 3.7.3 (picture taken from Galaktionov *et al.* (2003)). With each mixing element the number of material striations is doubled.

mixing and Bernoulli in the ergodic hierarchy, and we will make essentially no use of it in this book. Nevertheless, it is worth mentioning the idea behind it, even though there are technical issues.

First, 'K' stands for Kolmogorov, who invented this property (see Sinai (1989) for a discussion of the history), and it is based on a notion of *entropy*. Various notions of entropy play a major role in ergodic theory. Very roughly, the different notions of entropy quantify the degree of unpredictability of a system. But a proper discussion of the meaning of entropy in the context of mixing is beyond the scope of this book, and since we will not require it, we will omit a discussion of the K-property. Our work on showing that linked twist maps are mixing will be based on the Pesin theory of nonuniform hyperbolicity, whose use will allow us to conclude the stronger Bernoulli property. In fact, in Chernov & Haskell (1996) it is proved that nonuniformly hyperbolic K-systems are Bernoulli (this paper also contains a nice history of results on the Bernoulli property for various systems).

3.9 The Bernoulli property

The 'Bernoulli property' is a property that is stronger than mixing. In this section we will develop the framework in order to understand what it means for a map to have the Bernoulli property. A fair amount of background material from the subject of symbolic dynamics (the study of representing dynamics by shifts on symbol sequences) needs to be developed first.

As we have discussed, a dynamical system is a 'space', and a transformation that maps the space into itself. Here we will define our space to be the 'space of (bi-infinite) symbol sequences' (a good general text on symbol sequences and symbolic dynamics is Lind & Marcus (1995)). We can equip this space with a metric, and we can equip it with a σ-algebra and measure. Once we have done this, we will define a transformation of the space into itself (the 'shift', 'full shift' or 'Bernoulli shift') and we will derive some of its properties. After all of this, we will finally be at the point where we can say what it means for a map to have 'the Bernoulli property'. We include this construction for the simplest of dynamical systems in order to describe precisely what the Bernoulli property is, to show the complexity that a Bernoulli system possesses, and because it is instructive to see the mathematics that lies behind it. For the more complicated dynamical systems in later chapters we do not repeat this argument, but rather refer to theorems which have effectively extended the following ideas to a more general class of system.

3.9.1 The space of (bi-infinite) symbol sequences, Σ^N

Let $S = \{1, 2, 3, \dots, N\}$, $N \geq 2$ be our collection of symbols. We will build our sequences from elements of S, and the set of all such sequences comprises the space of all symbol sequences, which we will refer to as Σ^N. It is natural to consider Σ^N as a Cartesian product of infinitely many copies of S. This construction will allow us to draw some conclusions concerning the properties of Σ^N based only on our knowledge of S, the structure which we give to S, and topological theorems on infinite products.

Σ^N is a compact metric space

We now give some structure to S; specifically, we want to make S into a metric space, which can be done with the following metric, for $a, b \in S$:

$$d(a,b) \equiv \begin{cases} 1 & \text{if} \quad a \neq b, \\ 0 & \text{if} \quad a = b. \end{cases} \tag{3.2}$$

It is trivial to check that $d(\cdot, \cdot)$ is a metric (see Definition 3.2.1). The metric (3.2) actually induces the discrete topology on S, i.e., the topology defined by the collection of all subsets of S (see Munkres (1975)).

Since S consists of a finite number of points, it is trivial to verify that it is compact. Moreover, S is totally disconnected, i.e., its only connected subsets are one-point sets. We summarize the properties of S in the following proposition.

Proposition 3.9.1 *The set S equipped with the metric (3.2) is a compact, totally disconnected, metric space.*

We remark that compact metric spaces are automatically complete (i.e. containing all their limit points) metric spaces (see Munkres (1975), Section 7-3, Theorem 3.1).

Now we will construct Σ^N as a bi-infinite Cartesian product of copies of S:

$$\Sigma^N \equiv \cdots \times S \times S \times S \times S \times \cdots \equiv \prod_{i=-\infty}^{\infty} S^i \quad \text{where} \quad S^i = S \quad \forall i. \quad (3.3)$$

Thus, a point in Σ^N is represented as a 'bi-infinity-tuple' of elements of S:

$$s \in \Sigma^N \Rightarrow s = \{\ldots, s_{-n}, \ldots, s_{-1}, s_0, s_1, \ldots, s_n, \ldots\} \quad \text{where} \quad s_i \in S \quad \forall i,$$

or, more succinctly, we will write s as

$$s = \{\ldots s_{-n} \ldots s_{-1}.s_0 s_1 \ldots s_n \ldots\} \quad \text{where} \quad s_i \in S \quad \forall i.$$

A word should be said about the 'decimal point' that appears in each symbol sequence and has the effect of separating the symbol sequence into two parts, with both parts being infinite (hence the reason for the phrase 'bi-infinite sequence'). At present it does not play a major role in our discussion and could easily be left out with all of our results describing the structure of Σ^N going through just the same. In some sense, it serves as a starting point for constructing the sequences by giving us a natural way of subscripting each element of a sequence. This notation will prove convenient shortly when we define a metric on Σ^N. However, the real significance of the decimal point will become apparent when we define and discuss the shift map acting on Σ^N and its orbit structure.

In order to discuss limit processes in Σ^N, it will be convenient to define a metric on Σ^N. Since S is a metric space, it is also possible to define a metric on Σ^N. There are many possible choices for a metric on Σ^N; however, we will utilize the following. Consider the sequences:

$$s = \{\cdots s_{-n} \cdots s_{-1}.s_0 s_1 \cdots s_n \cdots\},$$

$$\bar{s} = \{\cdots \bar{s}_{-n} \cdots \bar{s}_{-1}.\bar{s}_0 \bar{s}_1 \cdots \bar{s}_n \cdots\} \in \Sigma^N.$$

The distance between s and $\bar{s}$ is defined as

$$d(s, \bar{s}) = \sum_{i=-\infty}^{\infty} \frac{1}{2^{|i|}} \frac{d_i(s_i, \bar{s}_i)}{1 + d_i(s_i, \bar{s}_i)}, \quad (3.4)$$

where $d_i(\cdot, \cdot)$ is the metric on $S^i \equiv S$ defined in (3.2). The reader should verify that (3.4) indeed defines a metric. Intuitively, this choice of metric implies that two symbol sequences are 'close' if they agree on a long central block. The following lemma makes this precise.

Lemma 3.9.1 *For $s, \bar{s} \in \Sigma^N$,*

(i) *Suppose $d(s, \bar{s}) < 1/(2^{M+1})$; then $s_i = \bar{s}_i$ for all $|i| \leq M$.*
(ii) *Suppose $s_i = \bar{s}_i$ for $|i| \leq M$; then $d(s, \bar{s}) \leq 1/(2^M)$.*

Proof The proof of (i) is by contradiction. Suppose the hypothesis of (i) holds and there exists some j with $|j| \leq M$ such that $s_j \neq \bar{s}_j$. Then there exists a term in the sum defining $d(s, \bar{s})$ of the form

$$\frac{1}{2^{|j|}} \frac{d_j(s_j, \bar{s}_j)}{1 + d_j(s_j, \bar{s}_j)}.$$

However, since $s_j \neq \bar{s}_j$,

$$\frac{d_j(s_j, \bar{s}_j)}{1 + d_j(s_j, \bar{s}_j)} = \frac{1}{2},$$

and each term in the sum defining $d(s, \bar{s})$ is positive so that we have

$$d(s, \bar{s}) \geq \frac{1}{2^{|j|}} \frac{d_j(s_j, \bar{s}_j)}{1 + d_j(s_j, \bar{s}_j)} = \frac{1}{2^{|j|+1}} \geq \frac{1}{2^{M+1}},$$

but this contradicts the hypothesis of (i).

We now prove (ii). If $s_i = \bar{s}_i$ for $|i| \leq M$, we have

$$d(s, \bar{s}) = \sum_{-\infty}^{i=-(M+1)} \frac{1}{2^{|i|}} \frac{d_i(s_i, \bar{s}_i)}{1 + d_i(s_i, \bar{s}_i)} + \sum_{i=M+1}^{\infty} \frac{1}{2^{|i|}} \frac{d_i(s_i, \bar{s}_i)}{1 + d_i(s_i, \bar{s}_i)};$$

however, $(d_i(s_i, \bar{s}_i)/(1 + d_i(s_i, \bar{s}_i))) \leq 1/2$ for all i, so we obtain

$$d(s, \bar{s}) \leq 2 \sum_{i=M+1}^{\infty} \frac{1}{2^{i+1}} = \frac{1}{2^M}.$$

$\square$

Armed with our metric, we can define neighbourhoods of points in Σ^N and describe limit processes. Suppose we are given a point

$$\bar{s} = \{\cdots \bar{s}_{-n} \cdots \bar{s}_{-1}.\bar{s}_0 \bar{s}_1 \cdots \bar{s}_n \cdots\} \in \Sigma^N, \quad \bar{s}_i \in S \quad \forall i, \tag{3.5}$$

and a positive real number $\varepsilon > 0$, and we wish to describe the 'ε-neighbourhood of $\bar{s}$', i.e., the set of $s \in \Sigma^N$ such that $d(s, \bar{s}) < \varepsilon$. Then, by Lemma 3.9.1, given $\varepsilon > 0$, we can find a positive integer $M = M(\varepsilon)$ such that $d(s, \bar{s}) < \varepsilon$ implies $s_i = \bar{s}_i \; \forall |i| \leq M$. Thus, our notation for an ε-neighbourhood of an arbitrary $\bar{s} \in \Sigma^N$ will be as follows

$$\mathcal{N}^{M(\varepsilon)}(\bar{s}) = \{s \in \Sigma^N | s_i = \bar{s}_i \quad \forall |i| \leq M, s_i, \bar{s}_i \in S \quad \forall i\}.$$

We are now ready to state our main theorem concerning the structure of Σ^N.

Proposition 3.9.2 *The space Σ^N equipped with the metric (3.4) is*

(i) *compact,*
(ii) *totally disconnected, and*
(iii) *perfect.*

Proof

(i) Since S is compact, Σ^N is compact by Tychonov's theorem (Munkres (1975), Section 5-1).
(ii) By Proposition 3.9.1, S is totally disconnected, and therefore Σ^N is totally disconnected, since the product of totally disconnected spaces is likewise totally disconnected (Dugundji (1966)).
(iii) Σ^N is closed, since it is a compact metric space. Let $\bar{s} \in \Sigma^N$ be an arbitrary point in Σ^N; then, to show that $\bar{s}$ is a limit point of Σ^N, we need only show that every neighbourhood of $\bar{s}$ contains a point $s \neq \bar{s}$ with $s \in \Sigma^N$. Let $\mathcal{N}^{M(\varepsilon)}(\bar{s})$ be a neighbourhood of $\bar{s}$ and let the symbol $\hat{s} = \bar{s}_{M(\varepsilon)+1} + 1$ if $\bar{s}_{M(\varepsilon)+1} \neq N$, and $\hat{s} = \bar{s}_{M(\varepsilon)+1} - 1$ if $\bar{s}_{M(\varepsilon)+1} = N$. Then the sequence

$$\{\cdots \bar{s}_{-M(\varepsilon)-2}\hat{s}\bar{s}_{-M(\varepsilon)}\cdots \bar{s}_{-1}.\bar{s}_0\bar{s}_1 \cdots \bar{s}_{M(\varepsilon)}\hat{s}\bar{s}_{M(\varepsilon)+2}\cdots\}$$

is contained in $\mathcal{N}^{M(\varepsilon)}(\bar{s})$ and is not equal to $\bar{s}$; thus Σ^N is perfect. □

We remark that the three properties of Σ^N stated in Proposition 3.9.2 are often taken as the defining properties of a *Cantor set*, of which the classical Cantor 'middle-thirds' set is a prime example. The following theorem of Cantor gives us information concerning the cardinality of perfect sets.

Theorem 3.9.1 *Every perfect set in a complete space has at least the cardinality of the continuum.*

Proof See Hausdorff (1962). □

Hence, Σ^N is uncountable.

Σ^N is a measure space

We are now going to provide some elements of the proof that Σ^N is a measure space. This may strike the reader as a bit abstract and unnecessary for the applications of mixing. Perhaps there is some truth to this. However, we will want to show that the shift map defined on Σ^N is measure-preserving and mixing, and this will require these concepts.

We need to equip Σ^N with a σ-algebra and then define a measure on the σ-algebra. We follow Arnold & Avez (1968), who construct a σ-algebra from the following sets:

$$A_i^j = \{s \in \Sigma^N | s_i = j, i \in \mathbb{Z}, j \in S\}. \tag{3.6}$$

In words, A_i^j is the set of bi-infinite symbol sequences that have j for the ith entry. These sets generate the σ-algebra in the sense, described earlier, that there exists a smallest σ-algebra containing these sets. From these sets we construct the following sets (known as 'cylinder sets'):

$$A_{i_1\ldots i_k}^{j_1\ldots j_k} \equiv A_{i_1}^{j_1} \cap \cdots \cap A_{i_k}^{j_k} = \{s \in \Sigma^N | s_{i_1} = j_1, \ldots, s_{i_k} = j_k\},$$

$$\text{where} \quad \{i_1, \ldots, i_k\} \quad \text{are all different.} \tag{3.7}$$

These are essentially all the sets in the σ-algebra on Σ^N in the sense that in order to verify properties such as measure preservation and mixing, it suffices to verify such properties just for these sets (see Katok & Hasselblatt (1995)).

Now we want to define a measure on Σ^N. This is done in three steps. First we define a normalized measure on S as follows:

$$\text{for } i \in S, \text{ we define } \mu(i) = p_i, \text{ where } p_i \geq 0, \text{ and } \sum_{i=1}^N p_i = 1.$$

Then we define the measure of the sets that generate the σ-algebra as:

$$\mu(A_i^j) = p_j. \tag{3.8}$$

Finally, we construct a measure on Σ^N by defining the measure of the cylinder sets:

$$\mu(A_{i_1}^{j_1} \cap \cdots \cap A_{i_k}^{j_k}) = \mu(\{s \in \Sigma^N | s_{i_1} = j_1, \ldots, s_{i_k} = j_k\}) = p_{j_1} \cdots p_{j_k}. \tag{3.9}$$

Of course, one needs to prove that the measure constructed in this way satisfies the axioms of a measure, but we leave this to the interested reader.

3.9.2 The shift map

Now that we have established the structure of Σ^N, we define a map on Σ^N, denoted by σ, as follows. For $s = \{\cdots s_{-n} \cdots s_{-1}.s_0 s_1 \cdots s_n \cdots\} \in \Sigma^N$, we define

$$\sigma(s) = \{\cdots s_{-n} \cdots s_{-1} s_0.s_1 \cdots s_n \cdots\},$$

or $[\sigma(s)]_i \equiv s_{i+1}$. The map σ is referred to as the *shift map,* and when the domain of σ is taken to be all of Σ^N, it is often referred to as a *full shift on N symbols.* We have the following proposition concerning some properties of σ.

Proposition 3.9.3 (i) $\sigma(\Sigma^N) = \Sigma^N$.

           (ii) σ *is continuous.*

Proof The proof of (i) is obvious. To prove (ii) we must show that, given $\varepsilon > 0$, there exists a $\delta(\varepsilon)$ such that $d(s, \bar{s}) < \delta$ implies $d(\sigma(s), \sigma(\bar{s})) < \varepsilon$ for $s, \bar{s} \in \Sigma^N$. Suppose $\varepsilon > 0$ is given; then choose M such that $1/(2^{M-1}) < \varepsilon$. If we then let $\delta = 1/2^{M+1}$, we see by Lemma 3.9.1 that $d(s, \bar{s}) < \delta$ implies $s_i = \bar{s}_i$ for $|i| \leq M$; hence, $[\sigma(s)]_i = [\sigma(\bar{s})]_i$, $|i| \leq M - 1$. Then, also by Lemma 3.9.1, we have $d(\sigma(s), \sigma(\bar{s})) < 1/2^{M-1} < \varepsilon$. $\square$

We now want to consider the orbit structure of σ acting on Σ^N. We have the following proposition.

Proposition 3.9.4 *The shift map σ has*

  (i) *a countable infinity of periodic orbits consisting of orbits of all periods;*
 (ii) *an uncountable infinity of nonperiodic orbits; and*
(iii) *a dense orbit.*

Proof (i) The proof is standard (Wiggins (2003)). In particular, the orbits of the periodic symbol sequences are periodic, and there is a countable infinity of such sequences. (ii) By Theorem 3.9.1 Σ^N is uncountable; thus, removing the countable infinity of periodic symbol sequences leaves an uncountable number of nonperiodic symbol sequences. Since the orbits of the nonperiodic sequences never repeat, this proves (ii). (iii) To prove this we form a symbol sequence by stringing together all possible symbol sequences of any finite length. The orbit of this sequence is dense in Σ^N since, by construction, some iterate of this symbol sequence will be arbitrarily close to any given symbol sequence in Σ^N. $\square$

Proposition 3.9.5 σ^N *is measure-preserving with respect to the measure on* Σ^N *constructed in the previous section.*

Proof First, we show that σ^N is measure-preserving on the A_i^j. It is easy see, by definition of the shift map and the definition of A_i^j in (3.6), that we have $\sigma(A_i^j) = A_{i-1}^j$, and therefore:

$$\mu(\sigma(A_i^j)) = \mu(A_{i-1}^j) = p_j = \mu(A_i^j).$$

Applying this result to the cylinder sets (3.7) gives:

$$\mu\left(\sigma\left(A_{i_1\dots i_k}^{j_1\dots j_k}\right)\right) = \mu\left(\sigma\left(A_{i_1}^{j_1}\cap\cdots\cap A_{i_k}^{j_k}\right)\right),$$

$$= \mu\left(\sigma\left(A_{i_1}^{j_1}\right)\cap\cdots\cap\sigma\left(A_{i_k}^{j_k}\right)\right),$$

$$= \mu\left(A_{i_1-1}^{j_1}\cap\cdots\cap A_{i_k-1}^{j_k}\right),$$

$$= p_{j_1}\cdots p_{j_k} = \mu\left(A_{i_1\dots i_k}^{j_1\dots j_k}\right)\quad\text{using (3.9).}\qquad(3.10)$$

$\square$

Proposition 3.9.6 σ *is mixing.*

Proof We need to verify Definition 3.7.1. We do this first for the sets A_i^j, A_n^m defined in (3.6), and note that $\sigma^n(A_i^j) = A_{i-n}^j$. The verification of Definition 3.7.1 is a straightforward calculation:

$$\mu(\sigma^n(A_i^j)\cap A_n^m) = \mu(A_{i-n}^j\cap A_n^m) = p_j p_m = \mu(A_i^j)\mu(A_n^m).$$

Next, we verify Definition 3.7.1 for the sets $A_{i_1\dots i_k}^{j_1\dots j_k}$, $A_{m_1\dots m_k}^{n_1\dots n_k}$ defined in (3.7), and note that, similar to above, $\sigma^n(A_{i_1\dots i_k}^{j_1\dots j_k}) = A_{i_1-n\dots i_k-n}^{j_1\dots j_k}$.

Again, the verification of Definition 3.7.1 is a straightforward calculation for these sets:

$$\mu\left(\sigma^n\left(A_{i_1\dots i_k}^{j_1\dots j_k}\right)\cap A_{m_1\dots m_k}^{n_1\dots n_k}\right) = \mu\left(A_{i_1-n\dots i_k-n}^{j_1\dots j_k}\cap A_{m_1\dots m_k}^{n_1\dots n_k}\right) = p_{j_1\dots j_k}p_{m_1\dots m_k}$$

$$= \mu\left(A_{i_1\dots i_k}^{j_1\dots j_k}\right)\mu\left(A_{m_1\dots m_k}^{n_1\dots n_k}\right).$$

$\square$

3.9.3 What it means for a map to have the Bernoulli property

Consider a map $f : M \to M$, and let $V \subset M$ denote an invariant set of f, i.e., $f(V) = V$ (V could possibly be M). Then f is said to have the Bernoulli property on V if it is 'isomorphic to a Bernoulli shift, mod 0'. More precisely, this means that the following diagram commutes:

$$
\begin{array}{ccc}
\Sigma^N & \xrightarrow{\ \sigma\ } & \Sigma^N \\
{\scriptstyle\phi}\downarrow & & \downarrow{\scriptstyle\phi} \\
V & \xrightarrow{\ f\ } & V
\end{array}
$$

where ϕ is an isomorphism. We can view ϕ as a 'change of coordinates' that transforms f, on the invariant set V to the Bernoulli shift, $\sigma = \phi^{-1}\circ f\circ\phi$.

This is significant because of the following result.

Proposition 3.9.7 *A map having the Bernoulli property is mixing.*

This is proven in Mané (1987). For us this is significant because Pesin's theory will enable us to deduce that linked twist maps are Bernoulli on a set of full measure, and therefore mixing.

Key point: A system enjoying the Bernoulli property is statistically indistinguishable from a sequence of random coin tosses, and as such is at the top of the ergodic hierarchy. In particular, showing that a system possesses the Bernoulli property is sufficient to show that the system also displays mixing behaviour.

As an example we now show that the Baker's transformation (recall Example 3.7.2) has the Bernoulli property.

The Baker's transformation has the Bernoulli property

Recall that an isomorphism is a one-to-one correspondence between points in the space of symbol sequences and points in the domain of the map that preserves the essential structures in both spaces. We construct it explicitly for the Baker's transformation B.

Every number in the unit interval $[0,1]$ can be represented by a binary expansion. In fact some points can be represented by two sequences, for example $1.000\ldots = 0.111\ldots$ in binary. These points however form a set of measure zero, and so can be ignored from a measure-theoretic point of view. Throughout this book sets of exceptional points with zero measure will be effectively disregarded. Here we can, by convention, use each 'half' of a bi-infinite sequence to form the binary expansion of a number in the unit interval as follows:

$$x = \sum_{i=0}^{\infty} \frac{s_i}{2^{i+1}}, \quad y = \sum_{i=1}^{\infty} \frac{s_{-i}}{2^i}.$$

This observation leads to a natural way for defining a map from Σ^2 to points in the unit square $U = [0,1] \times [0,1]$ as follows:

$$\phi(\{\cdots s_{-n}s_{-n+1}\cdots s_{-2}s_{-1}.s_0s_1s_2\cdots s_{n-1}s_n\cdots\})$$

$$= (x,y) \equiv \left(\sum_{i=0}^{\infty} \frac{s_i}{2^{i+1}}, \sum_{i=1}^{\infty} \frac{s_{-i}}{2^i} \right).$$

The goal now is to relate the shift dynamics on Σ^2 to the Baker's map on U through the map ϕ. This is related to what we mentioned in our (informal) definition of an isomorphism concerning 'preserving the essential structures'. Mathematically, this is accomplished by 'proving that the following diagram commutes'.

$$\begin{array}{ccc} \Sigma^2 & \xrightarrow{\sigma} & \Sigma^2 \\ \phi \downarrow & & \phi \downarrow \\ U & \xrightarrow{B} & U \end{array}$$

In other words, if we take any bi-infinite sequence of 0s and 1s, $s = \{\cdots s_{-n}s_{-n+1}\cdots s_{-2}s_{-1}.s_0 s_1 s_2 \cdots s_{n-1}s_n \cdots\}$, and we start in the upper left hand corner of the diagram, then map to U by first going across the top of the diagram, then down, we will get the same thing if we first go down, then across the bottom of the diagram. In other words, we want to show that for any sequence $s \in \Sigma^2$, we have $\phi \circ \sigma(s) = B \circ \phi(s)$. We will work out each side of the equality individually and, hence, show that they are equal.

From the definition of the map ϕ we have:

$$\phi \circ \sigma(s) = \left(\sum_{i=0}^{\infty} \frac{s_{i+1}}{2^{i+1}}, \sum_{i=1}^{\infty} \frac{s_{-i+1}}{2^i} \right).$$

This was the easy part. Showing the other side of the equality sign requires a bit more consideration. Note that the map is linear and diagonal on U. Therefore we can consider the Baker's map acting on the x component and y components individually.

For $x = \sum_{i=0}^{\infty} s_i/2^{i+1}$ we have $2x = \sum_{i=0}^{\infty} s_{i+1}/2^{i+1}$. For $y = \sum_{i=1}^{\infty} s_{-i}/2^i$ we have $(1/2)y = \sum_{i=1}^{\infty} s_{-i}/2^{i+1}$ (which is smaller than $1/2$), but this is not the end of the story. With respect to the y-component, we need to take into account the 'cutting and stacking' (i.e. the mod 1 part of the definition of the Baker's transformation). Note that in the expressions for $2x$ and $(1/2)y$ the symbol s_0 no longer appears. Somehow it needs to be put back in, and taking proper account of the cutting and stacking will do that. Now if $s_0 = 1$ the x component is greater than or equal to $1/2$. This means that after mapping by the Baker's transformation this point is outside the square, and therefore it is in the part that is cut off and stacked. Stacking means its y component is increased by $1/2$, i.e., $(1/2)y = \sum_{i=0}^{\infty} s_{-i}/2^{i+1}$. Now if $s_0 = 0$ this additional term contributes nothing to the sum. Combining these facts, we have shown

$$B \circ \phi(s) = B \left(\sum_{i=0}^{\infty} \frac{s_i}{2^{i+1}}, \sum_{i=1}^{\infty} \frac{s_{-i}}{2^i} \right) = \left(\sum_{i=0}^{\infty} \frac{s_{i+1}}{2^{i+1}}, \sum_{i=1}^{\infty} \frac{s_{-i+1}}{2^i} \right).$$

Now since ϕ is invertible $\phi \circ \sigma(s) = B \circ \phi(s)$ is equivalent to $B = \phi \circ \sigma \circ \phi^{-1}$. This looks very much like the formula for a similarity transformation for matrices.

We know that if two matrices are similar, they share many basic properties. The same is true in this more general setting. In particular, by composing this expression for B with itself n times we have $B^n = \phi \circ \sigma^n \circ \phi^{-1}$. From this we can conclude that there is a one-to-one correspondence between orbits of B and orbits of σ. In particular, we can then immediately conclude that B has an infinite number of (saddle-type) periodic orbits of all periods.

3.10 Summary

There is a rigorous heirarchy within ergodic theory. We have described above the main features of ergodicity, mixing and the Bernoulli property in detail, as these are the most immediately applicable to the problem of fluid mixing. Other than the K-property, which we have mentioned in Section 3.8, there are many other terms which can be found in ergodic theory textbooks (see Section 3.1 for a short list) which fit into this hierarchy (light mixing, mild mixing, partial mixing, mixing of order n, weakly Bernoulli – the list is seemingly endless!). However, these definitions are very technical and differences between them are probably not realizable in applications. The key features of the ergodic hierarchy can be summarized:

$$\text{Bernoulli} \implies K\text{-property} \implies (\text{strong}) \text{ mixing} \implies \text{ergodicity}.$$

4

Existence of a horseshoe for the linked twist map

The underlying structure of complicated behaviour in the linked twist map is that of the 'Smale horseshoe'. This chapter contains a detailed construction of the horseshoe, and the implications of its existence for symbolic dynamics.

4.1 Introduction

The main goal of the mathematical sections of this book is to show that linked twist maps have the Bernoulli property on all of their domain (except for possibly a set of measure zero). Before discussing the theory that will be necessary to attack this problem, we start with an easier, preliminary result. Namely, we give Devaney's proof of a theorem that a linked twist map has a Smale horseshoe (Devaney (1978)). This is a somewhat ambiguous, albeit commonly used statement in the literature. The Smale horseshoe map is a homeomorphism (it need not be area preserving) having the property that it has an invariant set on which the map is topologically conjugate to the Bernoulli shift, i.e., it has the Bernoulli property on an invariant set. A slight confusion may arise since occasionally the invariant set itself is referred to as *the horseshoe*. Smale horseshoe (or just "horseshoe") maps are ubiquitous in the sense that they can always be constructed near transverse homoclinic points. This is the content of the Smale–Birkhoff homoclinic theorem. All of this is described in detail, and from an elementary point of view, in Wiggins (2003).

From this description one might think that the theorem of Devaney that we will describe in this chapter will provide all of the results that we have said that we will prove in later chapters. After all, we have said that Devaney proved that the linked twist map has the Bernoulli property on an invariant set. The key here is the nature of this invariant set in Devaney's case. It is a Cantor set of measure zero, *not* the entire domain *except* for a set of measure zero. This is typical of the standard horseshoe construction. The 'chaotic invariant set' that arises from the construction is of measure zero. This raises the natural question of just how relevant it is to the dynamics since the probability of choosing an initial condition in this invariant set is zero. Certainly the horseshoe has an effect on an open neighbourhood of trajectories. However, quantifying this effect, or the size of the open neighbourhood, is generally quite difficult. This is what makes the results in later chapters so remarkable, and useful. One might think of them as 'horseshoe theorems on sets of full measure'. Nevertheless, this is an important result because it led Devaney to state that it is natural to conjecture that linked twist maps are ergodic. In this sense Devaney's result was a precursor to the later results on ergodicity, mixing, and the Bernoulli property on sets of full measure.

It is not our intention to dwell on the existence and properties of horseshoes in linked twist maps and their corresponding fluid flows. One reason for this is that horseshoes have been studied in great depth, in a variety of contexts, ranging from the very theoretical to the very practical, and it is not hard to find good expositions of the topic (see for example Guckenheimer & Holmes (1983), Ott (1993), Wiggins (2003), Lind & Marcus (1995)). More importantly however, this is a book about measure-theoretic aspects of dynamical systems – that is, properties which manifest themselves on sets of positive (or ideally, one set of full) measure. Horseshoes are fundamentally topological objects, and by their construction form a set of measure zero. There is no reason to believe, a priori, that the complexity inherent in a horseshoe should be shared by all (or almost all) points in a given domain. In Section 4.3 we see that this is indeed not the case, and refer to an experimental demonstration in which islands appear despite the presence of a horseshoe. Nevertheless, this chapter does have a place in this book, partly for completeness in our survey of classical linked twist map results, and partly because an understanding of the behaviour of the linked twist map which results in the horseshoe will help the understanding of later results.

4.2 The Smale horseshoe in dynamical systems

The Smale horseshoe was born out of a topologist's approach to dynamical systems. It is very similar to the Baker's map described in the previous chapter, except that whereas the Baker's map creates complexity via stretching and cutting, the horseshoe map creates complexity via stretching and folding. Like the Baker's map it is a diffeomorphism of the unit square, and produces a natural coding for symbolic dynamics. The classic reference is Smale (1967).

4.2.1 The standard horseshoe

A thorough description of the horseshoe map can be found in most textbooks on dynamical systems, for example Guckenheimer & Holmes (1983), Katok &

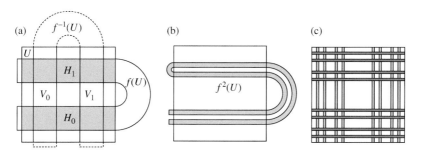

Figure 4.1 Illustration of the behaviour of the Smale horseshoe. In (a) we show the image $f(U)$ of the unit square U, which forms a horseshoe shape, and shade those parts which intersect U, giving a pair of horizontal strips H_0 and H_1. The preimages of H_0 and H_1 are the vertical strips V_0 and V_1 respectively. In (b) we show the second iterate of the map, which contains four horizontal strips, and (c) shows the intersections of horizontal and vertical strips at the third forward and backward iterate.

Hasselblatt (1995) and Robinson (1998), and so we give only a brief account. Let the unit square be denoted by $U = [0, 1] \times [0, 1]$. The action of the horseshoe map $f : U \to U$ is most easily defined with reference to Figure 4.1. The effect of applying f to U is to first stretch the square in the horizontal direction, then fold the image over in a horseshoe shape, and finally place the folded image back on top of the original U. As in Figure 4.1(a), some points fall outside U (and in fact almost all points iterate outside U eventually), but we consider only those points that remain in U by defining

$$H^{(1)} = H_0^{(1)} \cup H_1^{(1)} = f(U) \cap U,$$

and similarly, iterating the original square n times and keeping only those portions which remain in U, we double the number of horizontal strips at each forward iteration:

$$H^{(n)} = \bigcup_{i=0}^{i=n} H_{2^i-1}^{(n)} = f^n(U) \cap f^{n-1}(U) \cap \cdots \cap f(U) \cap U.$$

The second iterate of the map is shown in Figure 4.1(b), giving four horizontal strips. Since we are interested in finding a set which is invariant under forward and backward iterations, it is natural to ask where the horizontal strips $H^{(1)}$ came from. Note that it makes no sense to consider the preimage $f^{-1}(U)$ of the whole square U, as points in U map outside U, and similarly points in U have preimages outside U. However, the construction of the horseshoe is such that the preimages of the horizontal strips form vertical strips, as in Figure 4.1(a). In particular, we have

$$\begin{aligned} V_0^{(1)} &= f^{-1}(H_0^{(1)}) \\ V_1^{(1)} &= f^{-1}(H_1^{(1)}) \end{aligned}$$

and again, iterating further back we have

$$V_i^{(n)} = f^{-n}(H_i^{(n)}) \quad \text{for } i = 0, \ldots, 2^n - 1.$$

So a point which is initially in $V^{(n)}$ will arrive in the horizontal strips $H^{(n)}$ after n iterations. The invariant set Λ of the map must consist of points which remain in the horizontal strips $H^{(n)}$ under forward iteration, and the vertical strips $V^{(n)}$ under backward iteration, for each n. Therefore, Λ must be contained in the four squares formed by the intersection

$$H^{(1)} \cap V^{(1)} = (H_0^{(1)} \cup H_1^{(1)}) \cap (V_0^{(1)} \cup V_1^{(1)})$$

and also in the intersections $H^{(n)} \cap V^{(n)}$. The intersections for $n = 3$ are shown in 4.1(c). In fact the invariant set Λ is the intersection of a Cantor set of horizontal lines with a Cantor set of vertical lines,

$$\Lambda = \lim_{n \to \infty} H^{(n)} \cap V^{(n)}.$$

Just as in the Baker's map, we can specify a point by a bi-infinite sequence which records the location of the point at each iterate. Thus, for each $x \in \Lambda$, define a symbol sequence $s = \cdots s_{-2} s_{-1} \cdot s_0 s_1 s_2 \cdots$ of symbols $s_i \in S$, where

$$s_i = \begin{cases} 0 & \text{if } f^{-i}(x) \in H_0^{(1)}, \\[2mm] 1 & \text{if } f^{-i}(x) \in H_1^{(1)}. \end{cases}$$

Note that for a given n, $H_i^{(n)} \cap V_j^{(n)} \neq \emptyset$ for each $i, j \in [0, 2^n - 1]$ – that is, each horizontal strip intersects each vertical strip.

4.2.2 Symbolic dynamics

The dynamics on the invariant set for the horseshoe map can be expressed symbolically in a similar way to the Baker's map. Recall that in Section 3.9 we showed that the Baker's map possessed the Bernoulli property. The same conclusions can be drawn about the Smale horseshoe. In the literature, symbolic dynamics for horseshoe-type transformations are generally expressed in one of two representations. One is referred to as an *edge shift*, and the coding is given via an *adjacency matrix*. The other representation is a *vertex shift*, with the coding given via a *transition matrix*. These two methods contain essentially the same information, and indeed the definitions are frequently combined or interchanged, depending on the viewpoint of a particular author. We give a brief description of each.

Edge shift To create an edge shift we form a graph $\mathcal{G}^E$ composed of a set of n vertices $\mathcal{V}^E = \{V_1^E, \ldots, V_n^E\}$ and a set of m edges between vertices $\mathcal{E}^E = \{E_0^E, \ldots, E_{m-1}^E\}$. In this case the symbols in S are given by the edges, and the dynamics is represented by sequences of edges traversed in a bi-infinite walk on $\mathcal{G}^E$. An adjacency matrix $A = (a_{ij})$ encodes this information, and is given by

$$a_{ij} = \text{number of edges between vertex } V_i^E \text{ and vertex } V_j^E, 1 \leq i,j \leq n,$$

so that A is a non-negative $n \times n$ integer matrix. In the case of the Smale horseshoe, $\mathcal{G}^E$ takes a very simple form, shown in Figure 4.2(a). Noticing that the image $f(U)$ forms two horizontal strips H_0 and H_1 which span the width of U completely, we take a single vertex $\mathcal{V}^E = \{V_1^E\}$ corresponding to U and two edges $\mathcal{E}^E = \{E_0^E, E_1^E\}$, where E_i^E corresponds to the horizontal strip H_i. Thus A is a 1×1 matrix given by $A = (2)$. It can be shown that this symbolic system corresponds to the full shift on two symbols Σ_2 (essentially this is because either of the two edges E_i^E can follow each other), giving the Bernoulli property proven in the previous chapter for the Baker's map.

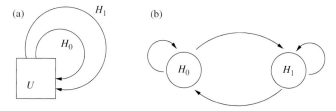

Figure 4.2 Two equivalent representations of the symbolic dynamics for the Smale horseshoe. We show (a) the graph $\mathcal{G}^E$ of the edge shift representation, and (b) the graph $\mathcal{G}^V$ of the vertex shift representation.

Vertex shift To create a vertex shift we form a graph $\mathcal{G}^V$ composed of a set of m vertices $\mathcal{V}^V = \{V_0^V, \ldots, V_{m-1}^V\}$ and a set of edges between vertices $\mathcal{E}^V = \{E_{ij}^V\}$. In this case the symbols in S are given by the vertices, and the dynamics is represented by the sequence of vertices visited by a bi-infinite walk on $\mathcal{G}^V$. A transition matrix $T = (t_{ij})$ encodes this information, and is given by

$$
t_{ij} = \begin{cases} 1 & \text{if there is an edge from vertex } V_i^V \text{ to } V_j^V \\ 0 & \text{if there is no edge from vertex } V_i^V \text{ to } V_j^V \end{cases}
$$

so that T is an $m \times m$ binary (i.e., consisting of 0s and 1s) matrix. We show the graph $\mathcal{G}^V$ for the Smale horseshoe in figure 4.2b). We have two vertices, representing the horizontal strips H_0 and H_1 (and hence the symbols 0 and 1), and edges representing the intersections $V_i \cap H_j$. Then T is a 2×2 matrix given by $T = \left(\begin{smallmatrix} 1 & 1 \\ 1 & 1 \end{smallmatrix}\right)$. The fact that T contains no zero entries means that there are no forbidden transitions in the symbolic dynamics and again it can be shown that this corresponds to the full shift on two symbols.

Robinson (1998) provides an algorithm for extracting a binary transition matrix from an adjacency matrix. For each edge E_j^E, let $b(E_j^E)$ be the beginning vertex of the edge, and let $e(E_j^E)$ be the end vertex. Then the entries of the transition matrix $T = (t_{ij})$ are

$$
t_{ij} = \begin{cases} 1 & \text{if } e(E_i^E) = b(E_j^E), \\ 0 & \text{if } e(E_i^E) \neq b(E_j^E). \end{cases}
$$

Subshifts of finite type The presence of a zero in a transition matrix indicates that there is a forbidden transition between particular symbols in a symbol sequence. In other words, certain pairs of symbols cannot occur adjacent to each other in a sequence. More precisely, if a transition matrix T has an entry $t_{ij} = 0$, then the symbol i may not be followed by the symbol j in any symbol sequence. Recalling the construction of the space of sequences of N symbols, Σ^N and the shift map σ in section 3.9, a *subshift* is the same shift on a subset of Σ^N, where the subset is described as all the symbol sequences not including a forbidden sequence. A subshift is of *finite type* if the number of forbidden sequences is finite. Since the forbidden sequences are given by the transition matrix T, we can denote a subshift of finite type by Σ_T^N. For more details on subshifts of finite type, see Lind & Marcus (1995).

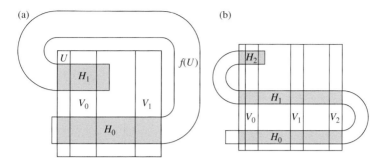

Figure 4.3 Two generalized horseshoes.

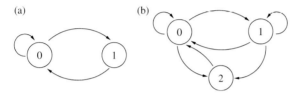

Figure 4.4 The graphs $\mathcal{G}^V$ for vertex shifts corresponding to the horseshoes in Figure 4.3.

4.2.3 Generalized horseshoes

The above has described the simplest type of horseshoe,[1] but there are many different ways to construct more general horseshoes to produce similar dynamical behaviour. Two such examples are illustrated in Figure 4.3. The first gives an example of an orientation preserving horseshoe with two strips, while the second gives an example of a horseshoe with three strips. Note that in both cases the image $f(U)$ does not produce strips which cover the entire width of U. For this reason we cannot construct an edge shift with a graph $\mathcal{G}^E$ using U as the only vertex in the manner of the above, although we can construct a vertex shift with a graph $\mathcal{G}^V$ and give a binary transition matrix as before. Figure 4.4(a) shows the graph $\mathcal{G}^V$ for the horseshoe in Figure 4.3(a), and Figure 4.4(b) shows the graph $\mathcal{G}^V$ for the horseshoe in Figure 4.3(b). It is straightforward to see that these horseshoes result in transition matrices T' and T'' respectively, given by

$$T' = \begin{pmatrix} 1 & 1 \\ 1 & 0 \end{pmatrix}, \qquad T'' = \begin{pmatrix} 1 & 1 & 1 \\ 1 & 1 & 1 \\ 1 & 0 & 0 \end{pmatrix}.$$

[1] We note that this horseshoe is orientation-reversing, and that arguably a orientation-preserving horseshoe is simpler.

and so these systems are conjugate to the subshifts of finite type $\Sigma_{T'}^2$ and $\Sigma_{T''}^3$ respectively.

A final comment about horseshoes in general is that they can be examples of *Anosov systems* – that is, they can be *uniformly hyperbolic*. Hyperbolicity will be discussed in detail in the following chapter, but for now we simply note that this property is a consequence of the stretching and contracting inherent in a horseshoe.

4.2.4 The Conley–Moser conditions

Typically a dynamical system will not have such a straightforward geometry as the horseshoe on the unit square described above, and hence symbolic dynamics may be difficult to construct directly. A set of conditions for verifying that a dynamical system contains a horseshoe was established in Moser (1973), and described in detail in Wiggins (2003). These have been generalized in Wiggins (1999). The *Conley–Moser* conditions are based on finding a foundation for Bakers map-like symbolic dynamics. We give a brief, heuristic description, based on Chien *et al.* (1986).

A transformation *f* contains a horseshoe if the following conditions hold.

1. There exists a quadrilateral Q such that the forward image $f(Q)$ forms
   'horizontal' strips which intersect Q and the backward image $f^{-1}(Q)$
   forms 'vertical' strips which intersect Q. In practice the strips need not lie
   horizontally and vertically. We will refer to the strips in this way (as does
   Moser (1973)) in order to make the connection with the theory for standard
   horseshoes, but any orientation will suffice, provided the strips are
   sufficiently transverse – that is, after a coordinate transformation to
   Cartesian coordinates, the angle between vertical and horizontal
   boundaries is sufficiently large. See Figure 4.5.

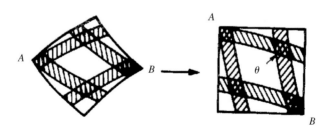

Figure 4.5 (Figure 5 from Chien *et al.* (1986)). The Conley–Moser condition for the intersection of the forward and inverse mappings.

2. The horizontal strips are the forward images of the vertical strips; moreover, the vertical boundaries of the vertical strips must map onto the vertical boundaries of the horizontal strips, and similarly for the horizontal boundaries.
3. The forward image of each horizontal strip must contain a new horizontal strip of width strictly less than the original strip. Similarly, vertical strips must contain thinner vertical images under backward iteration.

Although we have omitted all technical details from the Conley–Moser conditions, it is straightforward to see that they contain the essence of the Smale horseshoe construction above.

4.3 Horseshoes in fluids

Horseshoes began as mathematical constructs designed to demonstrate the existence of objects of great topological complexity in theoretical dynamical systems. They have since become regarded as a foundation of chaotic dynamics in many applications, to the extent that they have often been searched for numerically, and even experimentally. Fluid flows are one such field in which horseshoes may be observed, in diverse settings such as non-Newtonian cavity flows (Anderson *et al.* (2000)) and oceanography (Maas & Doelman (2001)). More examples of horseshoes in physical applications can be found in Moon (1987). Detailed descriptions of methods for locating horseshoes in fluid flows can be found in, for example, Ottino (1989a), Chien *et al.* (1986), Khakhar *et al.* (1986) and Ottino *et al.* (1994).

Such methods are generally based on the Conley–Moser conditions discussed above. The fact that these conditions can be verified for physical systems demonstrates that the horseshoe is not simply a theoretical construction, but actually manifests itself in practice. Chien *et al.* (1986) use experiments to locate horseshoes for a periodically alternating cavity flow, which is itself a form of linked twist map. Here in Figure 4.6 we reproduce Figure 10 from that paper, which shows an initial blob (a) and its image after forward iteration (b). In (c) the forward image is shown superimposed on a corresponding backward image. Finally in (d) the location of a quadrilateral is given which contains the horizontal and vertical strips required to satisfy the Conley–Moser conditions. However, as noted in Chien *et al.* (1986), the presence of a horseshoe does not guarantee good overall mixing, and some islands of unmixed fluid may persist.

Ottino *et al.* (1994) describes in more detail the challenges of experimentally verifying the Conley–Moser conditions for the blinking vortex flow, which was

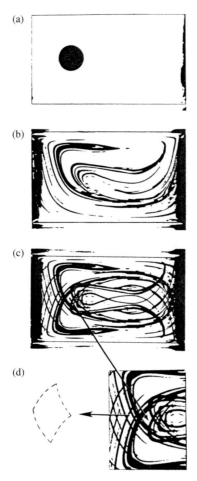

Figure 4.6 (Figure 10 from Chien *et al.* (1986)). The location of a quadrilateral containing horizontal and vertical strips satisfying the Conley–Moser conditions for a cavity flow.

discussed in Section 1.5.1 in the context of linked twist maps. They also discuss more general types of horsehoe in fluid flows. They conclude that the placement of the initial blob is crucial to the success of locating a horseshoe. For example, the initial blob must contain a periodic point; if not, the structure that develops may contain strips which fail to cover the entire width of the quadrilateral, violating the Conley–Moser conditions. The fact that it may be difficult to confirm the existence of a horseshoe experimentally makes it desirable to be

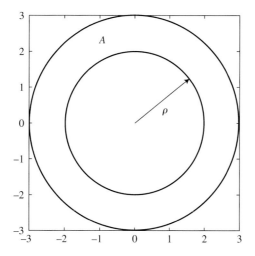

Figure 4.7 The annulus A, with inner radius ρ and outer radius 3.

able to prove the existence of such an object for a large class of systems, such as linked twist maps.

4.4 Linked twist mappings on the plane

First we define the linked twist mapping on the plane that we will study, using the notation of Devaney (1978). In order to do this, first we define a single-twist map.

4.4.1 A twist map on the plane

Let A denote the annulus in the plane, centred at the origin, with outer radius 3, inner radius ρ (see Figure 4.7). We define the following *twist map* $T : A \rightarrow A$ in standard polar coordinates (for which the counter-clockwise direction is the direction of increasing angle):

$$T(r, \theta) = \left(r, \theta + 2\pi \left(\frac{r - \rho}{3 - \rho}\right)\right). \tag{4.1}$$

The transformation T leaves the circles $r = $ constant ($\rho < r < 3$) invariant, rotating them through an angle $2\pi((r - \rho)/(3 - \rho))$. It is easy to see that the inner circle of A, $r = \rho$, is fixed by T, and the outer circle of A, $r = 3$, is rotated by an angle 2π, so that points on the outer circle are fixed (mod 2π). Moreover,

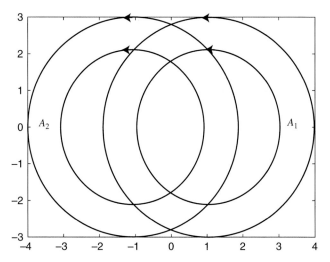

Figure 4.8 The intersecting annuli A_1 and A_2, centred at $(1,0)$ and $(-1,0)$ respectively.

it is clear that the angle of rotation of the circles between the inner and outer circles of the annulus increases monotonically with increasing r. Hence the name *twist map*.

4.4.2 Linking a pair of twist maps

We construct a *linked twist map* from two twist maps of the form of (4.1) as follows. Let A_1 and A_2 denote two copies of A centred at $(1,0)$ and $(-1,0)$ respectively (see Figure 4.8). Note that with this choice of ρ and annuli centres, $A_1 \cap A_2$ has exactly two disjoint components.

Let T_i denote the twist map on A_i, $i = 1, 2$. Since T_i is the identity map on the boundary of A_i, T_i can be extended to all of $A_1 \cup A_2$ by defining T_i to be the identity on $(A_1 \cup A_2) \backslash A_i$. We still refer to these maps defined on $A_1 \cup A_2$ in this way by T_i. Having extended the domains of T_i to $A_1 \cup A_2$ we can compose T_2 and T_1. Thus for integers j, $k \neq 0$ we define the resulting *linked twist map* on $A_1 \cup A_2$ as follows:

$$f_{j,k} = T_2^j \circ T_1^k : A_1 \cup A_2 \to A_1 \cup A_2. \tag{4.2}$$

In the remainder of this chapter we study the map $f_{j,k}$ for $j, k > 0$, with either $j > 1$ or $k > 1$. The analysis is similar in the case when either $j < 0$ or $k < 0$, but the case $|j| = |k| = 1$ must be considered separately for this particular construction.

4.5 Existence of a horseshoe in the linked twist map

We begin by stating Devaney's theorem of the existence of an invariant set on which $f_{j,k}$ is topologically conjugate to a subshift of finite type (i.e., existence of a horseshoe map).

Theorem 4.5.1 (Devaney (1978)) *There is a compact invariant hyperbolic set* $\Lambda_{j,k} \subset A_1 \cup A_2$ *on which* $f_{j,k}$ *is topologically conjugate to the subshift of finite type generated by the adjacency matrix*

$$\mathcal{P} = \begin{pmatrix} \alpha & \alpha - 1 \\ \alpha - 1 & \alpha \end{pmatrix}, \qquad (4.3)$$

where $\alpha = 2|j||k| - |j| - |k| + 1.$

4.5.1 Construction of the invariant set $\Lambda_{j,k}$

This construction will be for $j,\ k > 0$. Construction of the invariant set for other cases can be carried out with simple modifications of the argument to follow.

Let S_+ denote the component of $A_1 \cap A_2$ contained in the upper half plane, and let S_- denote the component contained in the lower half plane. We let A, B, C, and D denote the vertices of S_+, as shown in Figure 4.9. It should be clear that the sides AD and BC are fixed by T_1^k since, by definition, the inner

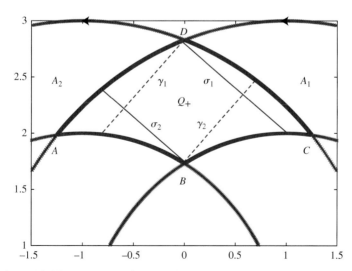

Figure 4.9 The component of $A_1 \cap A_2$ in the upper half plane, the vertices A, B, C, and D, and the curves γ_1, γ_2, σ_1, and σ_2.

boundary of A_1 is fixed by T_1 and the outer boundary of A_1 is rotated through an angle 2π. Similarly, the sides AB and DC are fixed by T_2^j.

Since the intersection of the annuli contains two components, S_+ and S_-, we will construct an edge shift (in the language of section 4.2.2) with two vertices. These vertices will be quadrilaterals Q_+ and Q_- contained in S_+ and S_- respectively, and correspond to the quadrilateral required for the Conley–Moser conditions. To make Q_+ and Q_- such that they satisfy these conditions we define the following curves.

Let $\gamma_1 \subset S_+$ denote a curve beginning at D and terminating at a point on AB so that it satisfies $T_1^k(\gamma_1) \subset CD$. This latter condition serves to define the termination point on AB, and hence to uniquely define γ_1. Note that the existence of such a unique curve follows from the monotone twist property, which is used in the construction of three additional curves. Let $\gamma_2 \subset S_+$ denote a curve beginning at B and terminating at a point on CD so that it satisfies $T_1^k(\gamma_2) \subset AB$. Let $\sigma_1 \subset S_+$ denote a curve beginning at D and terminating at a point on BC so that it satisfies $T_2^{-j}(\sigma_1) \subset AD$. Finally, let $\sigma_2 \subset S_+$ denote a curve beginning at B and terminating at a point on AD so that it satisfies $T_2^{-j}(\sigma_2) \subset BC$. These curves are shown in Figure 4.9. These curves form the boundaries of a quadrilateral Q_+ in S_+, and referring back to the Conley–Moser conditions and to the construction of edge shift graphs, it will shortly become clear that these curves have been chosen in order that horizontal and vertical strips cover the entire width of the quadrilateral, and do not merely intersect from one side, in the manner of the horseshoes of Figure 4.3.

By removing the ends of these curves, as shown in Figure 4.10, we denote the quadrilateral, Q_+, in S_+ with vertices given by $a = \gamma_1 \cap \sigma_2$, $c = \sigma_1 \cap \gamma_2$, $b = B$, and $d = D$. Using the same procedure, another quadrilateral, Q_-, can be constructed within S_-, with corresponding vertices denoted by a', b', c', and d'. It can also be obtained by rotating A_1 until Q_+ fits inside S_-.

As a final technical point before constructing an invariant set, we note that by choosing ρ close enough to 3 we can guarantee that $T_1^{-k}(ad) \cap S_- = \emptyset$, $T_1^{-k}(bc) \cap S_- = \emptyset$, $T_2^j(cd) \cap S_- = \emptyset$, and $T_2^j(ab) \cap S_- = \emptyset$, and similarly for the sides of Q_-.

The invariant set that we will construct is the following:

$$\Lambda_{j,k} = \bigcap_{n=-\infty}^{\infty} f_{j,k}^n(Q_+ \cup Q_-). \tag{4.4}$$

Certainly, by its very definition, this set is invariant (i.e., any point in this set remains in the set under all forward and backward iterations). It is also compact (i.e., for our purposes, closed and bounded). Perhaps it is not so obvious that it

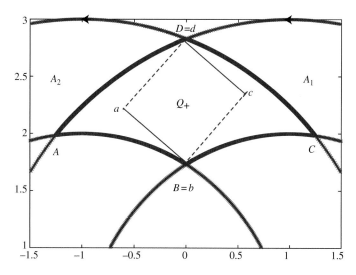

Figure 4.10 The quadrilateral Q_+.

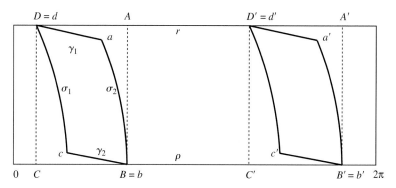

Figure 4.11 The annulus A_1 represented as a horizonal strip, periodic in the horizontal direction. The intersections S_+ and S_- of A_2 with the annulus are shown as dashed lines. The quadrilaterals Q_+ and Q_- are marked with solid lines.

is non-empty, and has the structure of a Cantor set of zero measure. This will follow from the inductive construction that we will describe.

Visualizing the action of $f_{j,k}$ on the quadrilaterals Q_+ and Q_- will be easier if we consider each annulus individually, cut it open and represent it as a rectangle with periodicity in the horizontal direction. This is shown for A_1 in Figure 4.11.

We begin by considering the action of T_1^k on Q_+. By the construction of Q_+ (i.e., the definitions of γ_1 and γ_2), under T_1^k the sides ad and bc of Q_+ are contracted and mapped into the boundary of A_2, while Q_+ is wrapped around

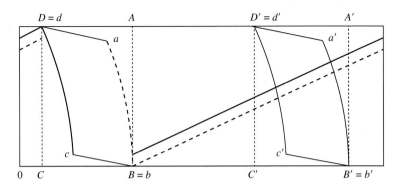

Figure 4.12 Action of T_1^k on Q_+ for $k = 1$. There are no intersections with Q_+, and one intersection with Q_-.

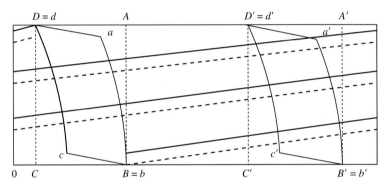

Figure 4.13 Action of T_1^k on Q_+ for $k = 3$. There are two disjoint components of intersection with Q_+, and three disjoint components of intersection with Q_-.

A_1 almost k full times. What 'almost' means is described in the figures below. We show the action of T_1^k on Q_+ in Figure 4.12 for $k = 1$, and in Figure 4.13 for $k = 3$. It should be clear that

$T_1^k(Q_+) \cap Q_+$ consists of $k - 1$ disjoint, parallel strips connecting ab to cd,

$T_1^k(Q_+) \cap Q_-$ consists of k disjoint, parallel strips connecting $a'b'$ to $c'd'$.

The same arguments can be applied to the action of T_1^k acting on Q_- to conclude that:

$T_1^k(Q_-) \cap Q_+$ consists of k disjoint, parallel strips connecting ab to cd,

$T_1^k(Q_-) \cap Q_-$ consists of $k - 1$ disjoint, parallel strips connecting $a'b'$ to $c'd'$.

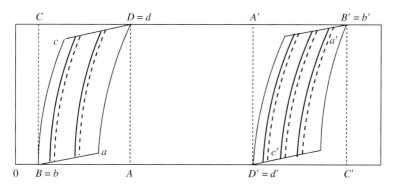

Figure 4.14 The strips in Q_+ and Q_- from Figure 4.13 with $k = 3$ shown in A_2.

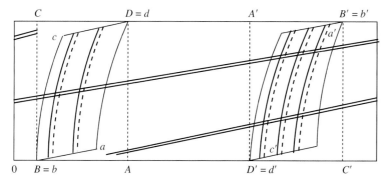

Figure 4.15 Intersection of $T_2^j(q_+)$ with Q_+ and Q_- in A_2 for $j = 2$.

In summary, the action of T_1^k on Q_+ and Q_- creates $k + (k-1) = 2k - 1$ strips in Q_+ and $2k - 1$ strips in Q_-. Now we represent the annulus A_2 as a rectangle in the same way, as we illustrate in Figure 4.14 with the strips discussed above shown for $k = 3$. Choose one of these strips (call it q_+) in $T_1^k(Q_+ \cup Q_-) \cap Q_+$, and consider the action of T_2^j on q_+. In the same manner as above, T_2^j wraps q_+ around A_2 almost j full times. In particular:

$T_2^j(q_+) \cap Q_+$ consists of $j - 1$ disjoint, parallel strips connecting ad to bc,

$T_2^j(q_+) \cap Q_-$ consists of j disjoint, parallel strips connecting $a'd'$ to $b'c'$.

This is shown in Figure 4.15 for $j = 2$. Recall that we have chosen ρ such that $T_2^j(cd) \cap S_- = \emptyset$, and $T_2^j(ab) \cap S_- = \emptyset$. This ensures that the image of q_+ 'begins and ends' outside the quadrilateral Q_-, so that in adherence with

Table 4.1. *Summary of the locations of images of Q_+ and Q_- after the application of T_1^k, and then T_2^j.*

Initial quadrilateral	after T_1^k	after T_2^j
Q_+	$k-1$ strips in Q_+	$(k-1)(j-1)$ strips in Q_+
		$j(k-1)$ strips in Q_-
	k strips in Q_-	jk strips in Q_+
		$k(j-1)$ strips in Q_-
Q_-	k strips in Q_+	$k(j-1)$ strips in Q_+
		kj strips in Q_-
	$k-1$ strips in Q_-	$j(k-1)$ strips in Q_+
		$(k-1)(j-1)$ strips in Q_-

the Conley–Moser conditions we have strips covering the entire width of the quadrilaterals.

Similarly, choose one of the strips in $T_1^k(Q_+ \cup Q_-) \cap Q_-$, (call it q_-) and consider the action of T_2^j on q_-. In the same manner as above, T_2^j wraps q_- around A_2 almost j times. In particular:

$T_2^j(q_-) \cap Q_+$ consists of j disjoint, parallel strips connecting *ad* to *bc*,

$T_2^j(q_-) \cap Q_-$ consists of $j-1$ disjoint, parallel strips connecting *a'd'* to *b'c'*.

Each of the strips formed by T_1^k is acted on in this way, and so we can give the combined effect of the linked twist map $f_{j,k} \equiv T_2^j T_1^k$ on each of Q_+ and Q_- as in Table 4.1.

Now consider $f_{j,k}(Q_+) \cap Q_+$ – that is, the intersection of the quadrilateral Q_+ with its image under $f_{j,k}$. Referring to Table 4.1 we see that it contains $(k-1)(j-1) + jk = 2kj - j - k + 1$ strips. Similarly, if we let

$$\alpha = 2jk - k - j + 1,$$

we can summarize our conclusions as follows:

$f_{j,k}(Q_+) \cap Q_+$ consists of α disjoint, parallel strips,

$f_{j,k}(Q_+) \cap Q_-$ consists of $\alpha - 1$ disjoint, parallel strips,

$f_{j,k}(Q_-) \cap Q_+$ consists of $\alpha - 1$ disjoint, parallel strips,

$f_{j,k}(Q_-) \cap Q_-$ consists of α disjoint, parallel strips.

Clearly, this procedure can be iterated, and

$$(Q_+ \cup Q_-) \cap f_{j,k}(Q_+ \cup Q_-) \cap f_{j,k}^2(Q_+ \cup Q_-) \cap \cdots \cap f_{j,k}^n(Q_+ \cup Q_-),$$

will consist of $(4\alpha - 2)^n$ strips, which get thinner as n increases, in line with the Conley–Moser conditions.

Next we need to consider $f_{j,k}^{-1}$, which is also a linked twist map. Proceeding in exactly the same manner as in the argument above,

$f_{j,k}^{-1}(Q_+) \cap Q_+$ consists of α disjoint, parallel strips connecting ab to cd,

$f_{j,k}^{-1}(Q_+) \cap Q_-$ consists of $\alpha - 1$ disjoint, parallel strips connecting $a'b'$ to $c'd'$,

$f_{j,k}^{-1}(Q_-) \cap Q_+$ consists of $\alpha - 1$ disjoint, parallel strips connecting ab to cd,

$f_{j,k}^{-1}(Q_-) \cap Q_-$ consists of α disjoint, parallel strips connecting $a'b'$ to $c'd'$.

We can also iterate this procedure:

$$(Q_+ \cup Q_-) \cap f_{j,k}^{-1}(Q_+ \cup Q_-) \cap f_{j,k}^{-2}(Q_+ \cup Q_-) \cap \cdots \cap f_{j,k}^{-n}(Q_+ \cup Q_-),$$

which gives $(4\alpha - 2)^n$ strips, which get thinner as n increases.

4.5.2 The subshift of finite type

Now we can construct an edge shift in the same manner as for the standard Smale horseshoe. We have two quadrilaterals Q_+ and Q_- and so the graph $\mathcal{G}_E$ has two vertices. The number of edges between vertices corresponds to the number of strips in the intersection of each quadrilateral with each image, as computed above. This results in the graph in Figure 4.16. The adjacency matrix $\mathcal{P} = \left(\begin{smallmatrix} \alpha & \alpha - 1 \\ \alpha - 1 & \alpha \end{smallmatrix} \right)$ from Theorem 4.5.1 is clearly the adjacency matrix

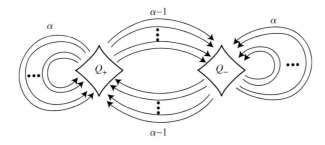

Figure 4.16 The graph that gives the symbolic dynamics of the invariant set $\Lambda_{j,k}$.

for the graph. We could follow the algorithm given earlier to construct a binary transition matrix for a vertex shift for this system. However, recall that the number of vertices in $\mathcal{G}_V$ is equal to the number of edges in $\mathcal{G}_E$, and so a transition matrix will be a $(4\alpha - 2) \times (4\alpha - 2)$ matrix of 0s and 1s.

4.5.3 The existence of the conjugacy

A conjugacy is a homeomorphism, ϕ, between $\Lambda_{j,k}$ and $\Sigma_{\mathcal{P}}^2$ having the property that $\sigma \circ \phi = \phi \circ f_{j,k}$. This latter property ensures that orbits of the subshift σ correspond to orbits of $f_{j,k}$, and vice versa. In Devaney (1978) it is stated that the conjugacy can be constructed using the methods in Moser (1973). This is essentially correct, but may be a bit confusing to the beginner as Moser (1973) does not deal with the case of subshifts. Nevertheless, with fairly easy modifications these methods can be adapted to the case of subshifts, and this is done in Wiggins (1988).

We will omit the details of the construction of the conjugacy here. It is instructive to thoroughly understand the construction of the conjugacy for the standard horseshoe map as described in, e.g., Wiggins (2003). Having done so, it should be clear that the method for this more complicated case follow the same pattern, although as is often the case, the details are intricate, and are rarely worked out in full.

A final point to make is that $\Lambda_{j,k}$ is a Cantor set. This requires the transition matrix to be irreducible, i.e., there exist some positive power of the matrix such that all entries are non-zero. This is only true if $|k|\,|j| > 1$, *for this particular construction of an invariant set*. In other words there may be other ways of constructing an invariant set that weaken this requirement.

Key point: The presence of a horseshoe in a linked twist map indicates the presence of complicated dynamics. However, a horseshoe is a set of zero volume, and so it is possible that the complexity may not be shared by a set of points of full or even positive volume.

4.5.4 Hyperbolicity of $\Lambda_{j,k}$

Finally, it must be shown that the invariant set is (uniformly) hyperbolic. Hyperbolicity considerations form a central part of the bulk of this book, and the proof that $\Lambda_{j,k}$ is hyperbolic is a consequence of work that is required later on. Therefore, we will refer back to this case in Chapter 6.

4.6 Summary

Following the construction in Devaney (1978) we have defined a linked twist map on the plane, and described an invariant set by analysing a carefully chosen pair of quadrilaterals in the intersection of the two linked annuli. The invariant set is given by the intersection of the images of the quadrilaterals under infinite forward and backward time. The way in which the image of each quadrilateral intersects the original pair of quadrilaterals defines a symbolic dynamics, which gives a measure of the complexity of the system. In later chapters we extend the symbolic dynamics from 'subshift of finite type', as in this construction, to 'full shift', and hence the Bernoulli property. Moreover, we will shortly describe how the result given in this chapter for a set of zero measure can be extended to a set of full measure.

5

Hyperbolicity

This chapter contains concepts and results from the field of hyperbolic dynamical systems. We define uniform and nonuniform hyperbolicity, and go on to describe Pesin theory, which creates a bridge between nonuniform hyperbolicity and the ergodic hierarchy.

5.1 Introduction

Hyperbolic dynamics, loosely speaking, concerns the study of systems which exhibit both expanding and contracting behaviour. Hyperbolicity is one of the most fundamental aspects of dynamical systems theory, both from the point of view of pure dynamical systems, in which it represents a widely studied and thoroughly understood class of system, and from the point of view of applied dynamical systems, in which it gives one of the simplest models of complex and chaotic dynamics. However the pay-off for this amount of knowledge and (apparent) simplicity is severe. While hyperbolic objects (for example certain fixed points and periodic orbits, and horseshoes, like that constructed in the previous chapter) are common enough occurrences, these are arguably of limited practical importance, as all these objects comprise sets of zero (Lebesgue) measure. There are only a handful of real systems for which the strongest form of hyperbolicity (uniform hyperbolicity) has been shown to exist on a set of positive measure. Typically, uniformly hyperbolic systems tend to be restricted to model systems, such as the Arnold Cat Map (Arnold & Avez (1968)), or idealized mechanical examples, such as the triple linkage of Hunt & Mackay (2003).

Weaker forms of hyperbolicity have been studied in great detail, and powerful results exist linking these to mixing properties, but still any sort of hyperbolicity is not a straightforward property to demonstrate. Moreover, despite its prevalence as a concept, definitions, results and conjectures pertaining to hyperbolicity in the literature can often be confusing, complicated, and even misleading. It is the intention of this chapter not to give a complete exposition of the theory of hyperbolic dynamical systems (such a project would probably require several volumes), but to give a clear and concise description of the ideas necessary for use in theorems on linked twist maps. In doing so we shall quote many seminal works, and refer to other illuminating texts.

We begin our discussion with the most restrictive type of hyperbolicity, namely uniform hyperbolicity, in which rates of expansion and contraction must have a constant bound that is valid for every trajectory. This very special type of behaviour was arguably first observed by Henri Poincaré around 1890 in the guise of homoclinic tangles (Poincaré (1890)). Much of the early inspiration behind hyperbolic dynamics came from geodesic flows, and the work of Jacques Hadamard, Pierre Duhem, George Birkhoff, Eberhard Hopf and Gustav Hedlund around the 1940s. See Hasselblatt (2002) for an excellent history of the subject. A systematic programme of study was initiated in the work of Dmitri Anosov, who was studying geodesic flows on surfaces of negative curvature (Anosov (1969)), and Stephen Smale, in the field of differential dynamics (Smale (1967)). The work that followed was a tremendous success, and mathematicians such as Anosov, Vladimir Arnold, Rufus Bowen, Yakov Sinai, David Ruelle and others advanced the understanding of uniformly hyperbolic systems. In particular Anosov diffeomorphisms, as they are also known, have been shown to be ergodic, mixing and to have the Bernoulli property. However, such systems are rare. Part of the problem with Anosov diffeomorphisms is that not every manifold can carry such a system – for example it is not possible for there to be an Anosov diffeomorphism on the circle or sphere (Katok & Hasselblatt 1995).

In view of the rarity of uniformly hyperbolic sets of positive measure appearing in applications, it is natural to wish to relax the stringent conditions. In 1977 Yakov Pesin published his seminal paper *Characteristic Lyapunov exponents and smooth ergodic theory* (Pesin (1977)), and created the field now known as *Pesin theory*, or the *theory of smooth nonuniformly hyperbolic dynamical systems*. In it he studied nonuniformly hyperbolic systems, having released the condition that growth rates must be uniform across the domain of a dynamical system. This paper establishes a rigorous link between nonuniform hyperbolicity and the non-vanishing of Lyapunov exponents. Further, it begins to construct the bridge between nonuniform hyperbolic systems and the ergodic hierarchy, specifically proving the result that the domain of any dynamical system displaying nonuniform hyperbolicity can be decomposed into a countable number of partitions on which the dynamics is ergodic. Also included are results on ergodicity and a Bernoulli decomposition. We give a brief summary of Pesin theory here, but note that a more thorough, and very accessible treatment is given in Pollicott (1983), and a recent account can be found in Barreira & Pesin (2002).

The link between nonuniform growth rates and Lyapunov exponents is a crucial one for applied mathematicians. Using Pesin's equivalent definition of nonuniform hyperbolicity as systems with non-zero Lyapunov exponents, we are faced with the challenge of formulating inequalities for infinite time limits rather than growth rates for every consecutive iteration of a map. This is still a non-trivial challenge, but the method of invariant cones discussed in Section 5.5 gives a well-known technique for extracting such exponents.

Also in this chapter we discuss some results from *Invariant Manifolds, Entropy and Billiards* due to Anatole Katok, Jean-Marie Strelcyn, François Ledrappier and Felix Przytycki (Katok *et al.* (1986)). This extends the results of Pesin theory to systems with singularities. A singularity in this case is defined in a precise way, and is different to the general notion usual in the applied mathematics literature (that of a property tending asymptotically to infinity). In smooth ergodic theory systems with singularities are systems which are not smooth at every point in the domain, but have a set of points at which the diffeomorphism loses some regularity. Provided such points are relatively uncommon, and the lack of smoothness not too severe, similar results concerning ergodic partitions and ergodicity can be formulated. Indeed Katok *et al.* (1986) goes further, and gives a global geometric condition to show the Bernoulli property (this is at least claimed as 'doubtless' in Katok *et al.* (1986)). This work was inspired by the study of billards, a favourite class of system for ergodic theorists, which typically have boundaries containing singularities. As we shall see in Chapter 6, linked twist maps can also be examples of smooth maps with singularities.

5.2 Hyperbolicity definitions

We begin by clarifying our use of notation in this chapter and the chapters to follow. We shall be interested in a transformation $f : M \to M$ (that is, a transformation f from a space M to itself). We discussed in Chapter 3 that M should be a compact metric space. Here we impose further restrictions on the structure of M. As will be found in textbooks on hyperbolicity, in general we require M to be a compact n-dimensional Riemannian C^∞ manifold. These properties were discussed in some detail in Chapter 3, but recall that a manifold is a topological space which can be viewed locally as Euclidean; a manifold is Riemannian if it is differentiable and has an inner product on tangent spaces; a Euclidean metric space is compact if it is closed and bounded. We will be interested solely in the torus and subsets of the plane, which are naturally compact Riemannian manifolds. For most of this chapter we will assume the transformation f is a smooth diffeomorphism (that is, infinitely times differentiable), but in Section 5.4 we relax this condition somewhat and discuss the case in which f is continuous everywhere but fails to be differentiable on a small subset of points (f is *smooth with singularities*).

For a diffeomorphism $f : M \to M$ we represent the derivative of f with respect to the variable x by $D_x f$ (this is the *Jacobian*). If no confusion can occur over which variable we are differentiating with respect to, we may use the shorthand Df. The notation $D_x f|_{x^*}$ refers to the derivative of f (with respect to x) evaluated at the point $x = x^*$. For each point $x \in M$, the *tangent space* is the space of all vectors tangent to M at x, and is written $T_x M$.

5.2.1 Uniform hyperbolicity

We begin by looking at the simplest type of dynamical behaviour, that of the fixed point. Recall that for a map $f : M \to M$ a fixed point is any point $x \in M$ for which $f(x) = x$.

Definition 5.2.1 (Hyperbolic fixed point) *A fixed point x^* of a map $f : M \to M$ is a hyperbolic fixed point if none of the eigenvalues of $D_x f|_{x^*}$ lie on the unit circle (that is, have modulus equal to one).*

Example 5.2.1 *Let $f : [0, 1] \to [0, 1]$, $f(x) = rx(1 - x)$ be the logistic map with $r \in [0, 4]$. This has fixed points at $x_1 = 0$ and $x_2 = 1 - (1/r)$ (if $r > 1$). The derivative is $D_x f = r(1 - 2x)$ and so $D_x f|_{x_1} = r$ and $D_x f|_{x_2} = 2 - r$. Hence x_1 is a hyperbolic fixed point unless $r = 1$, and x_2 is a hyperbolic fixed point unless $r = 3$ (recalling that x_2 exists only if $r > 1$).*

Example 5.2.2 *Let $g : \mathbb{R}^2 \to \mathbb{R}^2$, $g(x, y) = (\lambda x, \mu y)$ with $0 < \lambda < 1 < \mu$. The only fixed point is $(x^*, y^*) = (0, 0)$. Since Dg is the matrix*

$$Dg = \begin{pmatrix} \lambda & 0 \\ 0 & \mu \end{pmatrix}$$

we have eigenvalues λ and μ, and we have arranged that these do not lie on the unit circle. Therefore $(0, 0)$ is a hyperbolic fixed point. Figure 5.1 shows the behaviour of trajectories near $(0, 0)$. This illustrates the key idea of hyperbolicity – that of an expanding direction and a contracting direction (although definitions of hyperbolicity in the non-measure-preserving case may include the cases of entirely attracting or entirely repelling fixed points). We note here that if $\lambda \mu = 1$ (and so the map g is area-preserving), the curves in Figure 5.1 are the hyperbolæ $xy = $ const. This is the root of the term 'hyperbolic'.

Definition 5.2.1 can be easily extended to give a definition of a hyperbolic periodic point.

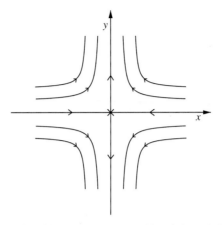

Figure 5.1 Trajectories of the linear map with uniformly hyperbolic fixed point of
Example 5.2.2. The fixed point $(0,0)$ is attracting along the x-axis and repelling
along the y-axis. Although this is a discrete time map we have drawn the trajectories
as continuous lines to illustrate the hyperbolæ which give hyperbolic systems
their name.

There is a key difference between Examples 5.2.1 and 5.2.2. In Example 5.2.2,
g is a linear map and so the derivative Dg is the same at each point (x, y). This
fact allows us to extend the definition of hyperbolicity from fixed points to the
whole domain in the case of a linear map.

Definition 5.2.2 *A linear map $g : \mathbb{R}^n \to \mathbb{R}^n$ is hyperbolic if none of the
eigenvalues of Dg lie on the unit circle.*

It will be useful to describe the dichotomy of expanding and contracting
directions in a more mathematical way. Consider a linear map $g : \mathbb{R}^n \to \mathbb{R}^n$
with eigenvalues λ_i, $i = 1, \ldots n$, which a priori can satisfy $|\lambda_i| < 1$, $|\lambda_i| = 1$
or $|\lambda_i| > 1$. Define a subspace E^s which is a subspace of $T\mathbb{R}^n$, and is the span
of all eigenvectors (or generalized eigenvectors) corresponding to $|\lambda_i| < 1$. We
term this the *stable* or *contracting* subspace and it contains all vectors which are
contracting under the linear map Dg. Similarly define an unstable or expanding subspace $E^u \subset T\mathbb{R}^n$ which is the span of all (generalized) eigenvectors
corresponding to $|\lambda_i| > 1$. The only remaining vectors are those which do not
have exponential expansion or contraction, but change at a slower rate, that is,
the (generalized) eigenvectors corresponding to $|\lambda_i| = 1$. We define a subspace
E^0 to contain such vectors. Having included all possible vectors we see that
$E^s \oplus E^u \oplus E^0 = \mathbb{R}^n$. This is called a *splitting* of the tangent space. An equivalent
definition of hyperbolicity is to require that $E^0 = \emptyset$, so that all tangent vectors
are either exponentially expanding or exponentially contracting.

In the more general case (that is, a map which is not linear) we cannot simply create a global splitting of the tangent space in the above manner. For example, referring to the logistic map in Example 5.2.1, we see that eigenvalues of $D_x f$ change depending on where the derivative is evaluated. Instead we must find a splitting of tangent space at *each* point x.

Definition 5.2.3 (Anosov diffeomorphism) *A diffeomorphism $f : M \to M$ of a compact Riemannian manifold M is* Anosov *if there exist constants $c > 0$, $0 < \lambda < 1$ and a continuous splitting of tangent space $T_x M = E_x^s \oplus E_x^u$ at each $x \in M$ such that*

$$(\textbf{A1}) \begin{cases} D_x f E_x^s = E_{f(x)}^s, & (5.1) \\ D_x f E_x^u = E_{f(x)}^u. & (5.2) \end{cases}$$

$$(\textbf{A2}) \begin{cases} \|D_x f^n v_s\| \le c\lambda^n \|v_s\| & \text{for } v_s \in E_x^s, & (5.3) \\ \|D_x f^{-n} v_u\| \le c\lambda^n \|v_u\| & \text{for } v_u \in E_x^u. & (5.4) \end{cases}$$

Condition **(A1)** states that E_x^s and E_x^u should be invariant under the action of the derivative Df. Thus stable subspaces map into stable subspaces, and unstable subspaces map into unstable subspaces as we iterate the map. Condition **(A2)** gives a rate of contraction for vectors $v \in E_x^s$ under forward iteration, and contraction for $v \in E_x^s$ under backward iteration. One often sees in definitions of Anosov diffeomorphisms Equation (5.4) given as

$$\|D_x f^n v_u\| \ge c^{-1} \lambda^{-n} \|v_u\| \qquad \text{for } v_u \in E_x^u \qquad (5.5)$$

This version of the definition makes it explicit that while forward iterations of the map contract vectors in E_x^s, the same iterations expand vectors in E_x^u. Definition 5.2.3 as given here appears the most commonly perhaps because, as remarked in Robinson (1998), under forward iterations any vector containing a contribution from E_x^u will be expanded exponentially, and not just vectors in E_x^u, and so forward iterates do not uniquely characterize the expanding subspace. However Equations (5.4) and (5.5) clearly express the same type of behaviour, as setting $w = D_x f^{-n} v$ in (5.4) gives $\|w\| \le c\lambda^n \|v\| = c\lambda^n \|D_x f^n w\|$ and so $\|D_x f^n w\| \ge c^{-1} \lambda^{-n} \|w\|$. Note that $D_x f w \in E_{f(x)}^u$ by condition **(A1)**. However these growth rates are expressed, they encapsulate the key property of hyperbolic systems – that under iteration some vectors are expanded and others are contracted, or equivalently that there are expanding directions under forward iteration and (different) expanding directions under backward iteration.

Example 5.2.3 *As in Example 5.2.2, let $g : \mathbb{R}^2 \to \mathbb{R}^2$, $g(x, y) = (\lambda x, \mu y)$ have $0 < \lambda < 1 < \mu$. In this case the subspaces E^s and E^u are simply the eigenvectors corresponding to λ and μ respectively – that is, $(1, 0)^T$ and $(0, 1)^T$.*

It is simple to verify both conditions (**A1**) *and* (**A2**) *of Definition 5.2.3. Note also that a vector such as* $(1, 1)^{\mathrm{T}}$ *expands under forward iteration, despite not belonging to* E^u.

The constants c and λ given above are the same for each point x (that is, are *uniform* constants). For this reason Anosov systems are also called *uniformly hyperbolic* systems. (In fact such systems are *uniformly completely hyperbolic* systems. The term 'completely' refers to the fact that the splitting $E^s_x \oplus E^u_x$ spans the entire tangent space $T_x M$. However this term is almost always omitted from given definitions, and often so is uniformly. 'Hyperbolic' has virtually become a synonym for 'Anosov'.)

The constant $c > 0$ can be taken to be $c = 1$ by a change of the norm, which then must vary with x. Such a norm is called an *adapted norm*, or *Lyapunov norm*. See for example (Robinson (1998), Chernov & Markarian (2003)). The fact that the splitting should be continuous (that is, that the subspaces E^s_x and E^u_x depend continuously on x) is sometimes given as a separate condition (**A3**). This is now for historical reasons, as the continuity actually follows from the first two conditions (Chernov & Markarian (2003), Brin & Stuck (2002)).

Stable and unstable manifolds

Stable and unstable manifolds form the bedrock of hyperbolic behaviour as they govern the behaviour of the dynamics starting in the immediate vicinity of a given periodic point or trajectory. We begin with the standard definition of a local stable manifold[1] of a fixed point x^*, which defines for a neighbourhood of x^* the set of points which remain in that neighbourhood and lead to x^* under forward iteration. See for example Guckenheimer & Holmes (1983).

Definition 5.2.4 (local stable manifold) *The local stable manifold of a fixed point x^* of a (C^1) diffeomorphism $f : M \to M$ in a neighbourhood $B(x^*, r)$ of x^* is given by*

$$\gamma^s(x^*) = \{x \in B(x^*, r) | d(f^n(x), x^*) \to 0 \text{ as } n \to \infty\}$$

where $B(x^*, r)$ is the open ϵ-ball of radius r about x^* defined in Definition 3.2.5.

Since we are concerned with diffeomorphisms, which are by definition invertible, we can define a local unstable manifold in the same way, looking at backward iterates:

Definition 5.2.5 (local unstable manifold) *The local unstable manifold of a fixed point x^* of a (C^1) diffeomorphism $f : M \to M$ in a neighbourhood*

---

[1] Strictly speaking, Definition 5.2.4 defines a local stable *set*. The fact that it is also a manifold comes from Theorem 5.2.1.

$B(x^*, r)$ *of* x^* *is given by*

$$\gamma^u(x^*) = \{x \in B(x^*, r) | d(f^{-n}(x), x^*) \to 0 \text{ as } n \to \infty\}$$

The local unstable manifold may also be defined for non-invertible maps but that will not concern us here (see (Robinson (1998))). The question of whether such sets exist for a given fixed point, and what form they take, is answered by the classical *stable manifold theorem for fixed points*.

Theorem 5.2.1 (Stable manifold theorem for fixed points) *Let* $f : M \to M$ *be a diffeomorphism, and let* x^* *be a hyperbolic fixed point. Then there exist local stable and unstable manifolds* $\gamma^s(x^*)$ *and* $\gamma^u(x^*)$ *of the same dimension as, and tangent to, the subspaces* $E_{x^*}^s$ *and* $E_{x^*}^u$ *respectively. The local manifolds* $\gamma^s(x^*)$ *and* $\gamma^u(x^*)$ *are as smooth as* f *(so in particular, they are differentiable, since* f *is a diffeomorphism). The size of the local stable and unstable manifolds is given by the radius* r *of the neighbourhood* $B(x^*, r)$.

These ideas can be extended throughout the domain of an Anosov diffeomorphism to give stable and unstable manifolds at each point.

Theorem 5.2.2 (Hadamard–Perron stable manifold theorem) *Let* $f : M \to M$ *be an Anosov diffeomorphism. Then for each* $x \in M$ *there exist local stable and unstable manifolds*

$$\gamma^s(x) = \{y \in B(x, r(x)) | d(f^n(y), f^n(x)) \to 0 \text{ as } n \to \infty\}$$

$$\gamma^u(x) = \{y \in B(x, r(x)) | d(f^{-n}(y), f^{-n}(x)) \to 0 \text{ as } n \to \infty\}$$

of the same dimension as, and tangent to, the subspaces E_x^s *and* E_x^u *respectively. The local manifolds* $\gamma^s(x)$ *and* $\gamma^u(x)$ *are as smooth as* f *(so in particular, they are differentiable, since* f *is a diffeomorphism). The size of the local stable and unstable manifolds at* x *is given by the radius* $r(x)$ *of the neighbourhood* $B(x, r(x))$, *and moreover there is a uniform bound* $r \geq r' > 0$ *for all* $x \in M$.

Having verified the existence of local manifolds, we define global stable and unstable manifolds by taking unions of backward and forward iterates of local stable and unstable manifolds.

Definition 5.2.6 (global stable/unstable manifolds) *The global stable and unstable manifolds at each* $x \in M$, $W^s(x)$ *and* $W^u(x)$ *respectively, are given by*

$$W^s(x) = \bigcup_{n \geq 0} f^{-n}(\gamma^s(f^n(x))). \tag{5.6}$$

$$W^u(x) = \bigcup_{n \geq 0} f^n(\gamma^u(f^{-n}(x))). \tag{5.7}$$

We now give a straightforward, but useful, result which shows that forward time averages of continuous functions for a given point are equal to forward time averages of that function for any point in the stable manifold of that point. Recall Equation (3.1) for the definition of the forward time average of a function.

Lemma 5.2.1 *Let $f : M \to M$ be an Anosov diffeomorphism, and let $\phi : M \to \mathbb{R}$ be a continuous function. For any $x \in M$, let $y \in \gamma^s(x)$. Then $\phi^+(y) = \phi^+(x)$. Similarly, for any $x' \in M$, let $y' \in \gamma^u(x')$. Then $\phi^-(y') = \phi^-(x')$.*

Proof By the continuity of ϕ, for any fixed $\epsilon > 0$ there exists a $\delta > 0$ such that $|\phi(x) - \phi(x')| < \epsilon$ whenever $d(x, x') < \delta$. By Theorem 5.2.2 we can choose m such that $d(f^i(y), f^i(x)) < \delta$ whenever $i \geq m$. Now for $n > m$ we have

$$\frac{1}{n}\sum_{i=0}^{n-1}\phi(f^i(y)) - \frac{1}{n}\sum_{i=0}^{n-1}\phi(f^i(x)) = \frac{1}{n}\sum_{i=0}^{m-1}(\phi(f^i(y)) - \phi(f^i(x)))$$

$$+ \frac{1}{n}\sum_{i=m}^{n-1}(\phi(f^i(y)) - \phi(f^i(x))).$$

Hence by the triangle inequality,

$$\left|\frac{1}{n}\sum_{i=0}^{n-1}\phi(f^i(y)) - \frac{1}{n}\sum_{i=0}^{n-1}\phi(f^i(x))\right| \leq \frac{1}{n}\left|\sum_{i=0}^{m-1}(\phi(f^i(y)) - \phi(f^i(x)))\right| + \epsilon$$

and so since ϵ is arbitrary and the sum after the inequality is a finite sum we have

$$\lim_{n\to\infty}\left|\frac{1}{n}\sum_{i=0}^{n-1}\phi(f^i(y)) - \frac{1}{n}\sum_{i=0}^{n-1}\phi(f^i(x))\right| = 0.$$

It follows that $\phi^+(y) = \phi^+(x)$. An identical argument using backward iteration can be applied to show that $\phi^-(y') = \phi^-(x')$ for each $y' \in \gamma^u(x')$. $\square$

Results for uniform hyperbolicity

Anosov diffeomorphisms are among the best understood classes of dynamical system. Moreover, they are closely linked to the ergodic hierarchy. A thorough discussion can be found in Hasselblatt (2002). Briefly, geodesic flows of negative curvature (closely linked to Anosov systems) were much studied in the early 1900s. These were shown to be ergodic by Hopf (an excellent description of this result can be found in Brin & Stuck (2002)), and to be mixing by

Hedlund. When Anosov defined the class of systems which carry his name, he built on this work and showed that these systems possess the K-property (recall Section 3.8). The Bernoulli property for Anosov systems was shown in 1973 by Ornstein and Weiss for uniformly hyperbolic geodesic flows (Ornstein & Weiss (1973)), a result later generalized to more general Anosov systems (Ornstein & Weiss (1998)).

> **Key point:** Uniformly hyperbolic systems are a well-understood class of dynamical system, and in particular they enjoy the Bernoulli property. Their definition is based on uniform bounds for rates of expansion and contraction. However, they are rare in applications. Linked twist maps possess hyperbolic behaviour, but in a weaker form, which requires more work to draw similar conclusions.

Anosov diffeomorphisms are named after Dmitri Anosov. Anosov himself termed such systems *systems with a condition*, which has led them to be called C-systems, or sometimes Y- or U-systems (from a transliteration of the Russian for 'condition'). We will refer to Equations (5.3), (5.4) (and (5.5)) as *Anosov growth conditions*.

The Arnold Cat Map

The hyperbolic toral automorphism now widely known as the Arnold Cat Map is a canonical example of an Anosov map (see Arnold & Avez (1968) for more details, and Hasselblatt (2002) for some intriguing history). We have already seen numerically the behaviour of this map, but we will restate it here, and then use it throughout this chapter to illustrate the results and ideas to follow. It is of particular interest to us as linked twist maps on the torus are in some sense a generalized nonuniform version of the Cat Map.

Example 5.2.4 (Arnold Cat Map) *Let $f : \mathbb{T}^2 \to \mathbb{T}^2$ be an invertible map on the standard 2-torus given by $f(x, y) = (x + y, x + 2y)$. The action of the Arnold Cat Map is shown in Figure 5.2. The Jacobian $D_x f$ is given by the matrix*

$$D_x f = \begin{pmatrix} 1 & 1 \\ 1 & 2 \end{pmatrix}.$$

The Jacobian is identical for all $(x, y) \in \mathbb{T}^2$ so we will refer to it as Df. The determinate of Df, $|\det Df| = 1$ and so f is area-preserving and invertible. The eigenvalues of Df are $\lambda_{\pm} = 3/2 \pm \sqrt{5}/2$, with eigenvectors $v_{\pm} = (1, 1/2 \pm \sqrt{5}/2)$ respectively. As in Example 5.2.3 these eigenvectors give the stable

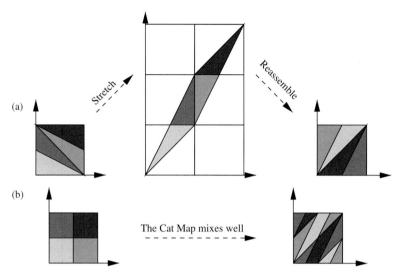

Figure 5.2 Two illustrations of the action of the Cat Map of Example 5.2.4. In (a) the unit square is first shaded in four parts as shown. The action of the Cat Map is to stretch and squeeze the unit square into the parallelogram in the centre diagram. Since the Cat Map is a toral map the shaded parts are re-assembled into the unit square. In (b) we illustrate the mixing qualities of the Cat Map. Here the initial unit square is shaded into quadrants, and their images shaded the same after one iterate of the map. It is clear that even after a single iterate the shaded areas are becoming mixed.

and unstable subspaces E^s and E^u. Moreover, the Arnold Cat Map has one-dimensional stable and unstable manifolds lying the direction of E^s and E^u respectively.

Key point: The Arnold Cat Map is a canonical example of a uniformly hyperbolic system, and its dynamics forms a skeleton of the more intricate dynamics in linked twist maps.

Uniformly hyperbolic trajectories

Uniform hyperbolicity is rare in applications, and in general difficult to establish. This is because the constants c and λ are required to be valid *for all* points in the domain. Often we may have good information about certain points, but less about others. For example, we may know much about points in the support of a certain invariant measure, such as periodic orbits. Thus one frequently sees definitions of (uniform) hyperbolicity for trajectories.

Definition 5.2.7 (uniformly (completely) hyperbolic trajectory) *A traject-ory[2] $\{f^m x\}$ of a diffeomorphism $f : M \to M$ of a compact Riemannian manifold is uniformly (completely) hyperbolic if there exist constants $c > 0$, $0 < \lambda < 1 < \mu$ and a continuous splitting of tangent space $T_{f^m x} M = E^s_{f^m x} \oplus E^u_{f^m x}$ such that for each $k \in \mathbb{Z}$ and $n > 0$,*

$$(\textbf{UHT1}) \begin{cases} D_x f^k E^s_x = E^s_{f^k(x)}, & (5.8) \\ D_x f^k E^u_x = E^u_{f^k(x)}. & (5.9) \end{cases}$$

$$(\textbf{UHT2}) \begin{cases} \|D_x f^n v_s\| \leq c \lambda^n \|v_s\| & \text{for } v_s \in E^s_{f^k(x)}, & (5.10) \\ \|D_x f^n v_u\| \geq c^{-1} \mu^n \|v_u\| & \text{for } v_u \in E^u_{f^k(x)}. & (5.11) \end{cases}$$

$$(\textbf{UHT3}) \angle (E^s_{f^k(x)}, E^u_{f^k(x)}) \geq c^{-1}. \qquad (5.12)$$

From this we can recover the definition of Anosov by demanding that all trajectories of f are uniformly hyperbolic with the same constants c, λ and μ, and replacing $\min\{\lambda, \mu^{-1}\}$ with λ. In condition (**UHT3**) the symbol '$\angle$' denotes the angle between the stable and unstable subspaces, and so this condition guarantees that these subspaces do not approach each other. Note that in this definition forward iterations are used to define growth along the orbit in the unstable subspace.

The difficulty of establishing uniform hyperbolicity makes it natural to weaken the very strict conditions (**A1**) and (**A2**) of Definition 5.2.3. This is done in two main ways – the uniformity of the constants c and λ may be relaxed to give nonuniform hyperbolicity, or the completeness of the splitting of tangent space may be relaxed to give partial hyperbolicity.

5.2.2 Nonuniform hyperbolicity

The idea of nonuniform hyperbolicity – that growth rates of tangent vectors should vary from point to point – is simple enough, but it requires precise mathematical statements to express it accurately. We begin by generalizing the definition of uniform hyperbolicity for a trajectory to nonuniform hyperbolicity for a trajectory.

Definition 5.2.8 (nonuniformly (completely) hyperbolic trajectory) *A tra-jectory $\{f^m_x\}$ is nonuniformly (completely) hyperbolic if there are numbers*

---

[2] The usual terminology for a trajectory, or orbit, is $\{f^m(x)\}$ or $\{f^m x\}$. We will stick to this use, except in subscripts when we may refer to a trajectory as simply $f^m x$.

$0 < \lambda < 1 < \mu$ *and a splitting* $T_{f^m(x)}M = E^s_{f^m(x)} \oplus E^u_{f^m(x)}$, *and if for all sufficiently small* $\epsilon > 0$ *there exists a function* $c(x, \epsilon)$ *such that for each* $k \in \mathbb{Z}$ *and* $n > 0$:

$$
\textbf{(NUHT1)} \quad
\begin{cases}
D_x f^k E^s_x = E^s_{f^k(x)}, & (5.13) \\
D_x f^k E^u_x = E^u_{f^k(x)}. & (5.14)
\end{cases}
$$

$$
\textbf{(NUHT2)} \quad
\begin{cases}
\|D_x f^n v\| \le c(f^{n+k}(x), \epsilon) \lambda^n \|v\| & \text{for } v \in E^s_{f^k(x)}, \quad (5.15) \\
\|D_x f^n v\| \ge c^{-1}(f^{n+k}(x), \epsilon) \mu^n \|v\| & \text{for } v \in E^u_{f^k(x)}. \quad (5.16)
\end{cases}
$$

$$\textbf{(NUHT3)} \quad \angle(E^s_{f^k(x)}, E^u_{f^k(x)}) \ge c^{-1}(f^k(x), \epsilon).$$

$$\textbf{(NUHT4)} \quad c(f^k(x), \epsilon) \le e^{\epsilon|k|} c(x, \epsilon).$$

Condition **(NUHT1)** is identical to condition **(A1)** (and **(UHT1)**). Conditions **(NUHT2)** and **(NUHT3)** are also similar to their counterparts from uniform hyperbolicity except that constants have been replaced by functions which depend on the location on the trajectory. Condition **(NUHT4)** is the crucial, technical part of this definition. It says that while the estimates on growth in conditions **(NUHT2)** may get worse along a trajectory, they do so relatively slowly.

To get the definition of nonuniform hyperbolicity for a diffeomorphism we must recognise that not only the constant c can vary along a trajectory, but also the constants λ, μ and ϵ may be different for different trajectories. Thus we replace such constants by functions $\lambda(x)$, $\mu(x)$ and $\epsilon(x)$.

Definition 5.2.9 (nonuniformly (completely) hyperbolic) *A diffeomorphism* $f : M \to M$ *is* nonuniformly (completely) hyperbolic *if there are (measurable) functions* $0 < \lambda(x) < 1 < \mu(x)$, *and* $\epsilon(x)$ *such that* $\epsilon(f^k(x)) = \epsilon(x)$ *(that is,* ϵ *is invariant, or constant, along each trajectory) and a splitting* $T_x M = E^s_x \oplus E^u_x$ *for each* x, *and a function* $c(x)$ *such that for every* $k \in \mathbb{Z}$ *and* $n > 0$:

$$
\textbf{(NUH1)} \quad
\begin{cases}
D_x f^k E^s_x = E^s_{f^k(x)}, & (5.17) \\
D_x f^k E^u_x = E^u_{f^k(x)}. & (5.18)
\end{cases}
$$

$$
\textbf{(NUH2)} \quad
\begin{cases}
\|D_x f^n v\| \le c(f^k(x)) \lambda^n(x) \|v\| & \text{for } v \in E^s_x, \quad (5.19) \\
\|D_x f^n v\| \ge c^{-1}(f^k(x)) \mu^n(x) \|v\| & \text{for } v \in E^u_x. \quad (5.20)
\end{cases}
$$

$$\textbf{(NUH3)} \quad \angle(E^s_x, E^u_x) \ge c^{-1}(x).$$

$$\textbf{(NUH4)} \quad c(f^k(x)) \le c(x) e^{\epsilon(x)|k|}.$$

> **Key point:** Nonuniform hyperbolicity is a loosening of the strict conditions of uniform hyperbolicity. Expanding and contracting behaviour is still present, but now the rate at which such behaviour occurs can vary from point to point. This remains a difficult property to demonstrate in applications, but we will see that linked twist maps form a paradigm example of a nonuniformly hyperbolic system.

5.2.3 Partial nonuniform hyperbolicity

Hyperbolicity can be weakened further by removing the necessity for linearly expanding and contracting directions to span the whole tangent space. In this case the hyperbolicity is not complete, but partial. Although we do not use this concept in this book, we mention the definition for thoroughness. This can be given simply for a trajectory by replacing the numbers $0 < \lambda < 1 < \mu$ with numbers $0 < \lambda < \min(1, \mu)$ so that μ is not necessarily greater than 1. The rest of the definition is the same. Similarly for nonuniform partial hyperbolicity of a diffeomorphism we require functions $\lambda(x)$, $\mu(x)$ and $\epsilon(x)$ such that $0 < \lambda(x) < \mu(x)$, $\mu(x) - \lambda(x) > \epsilon(x)$ and $1 - \lambda(x) > \epsilon(x) > 0$.

Since E^s and E^u do not generate the whole tangent space this is the situation in which $E^0 \neq \emptyset$, that is, there are directions which are not linearly expanding or contracting. The subspace E^0 is called neutral, and vectors in it may contract or expand but not exponentially. Finally we note that setting $\epsilon = 0$ gives the little-used concept of partial uniform hyperbolicity. See for example Brin & Pesin (1974) and Barreira & Pesin (2002) for more information.

5.2.4 Other hyperbolicity definitions

Frequently one sees definitions for hyperbolic sets. Thus

Definition 5.2.10 (uniformly hyperbolic set) *Let $f : M \to M$ be a diffeomorphism of a compact smooth Riemannian manifold M. An f-invariant set A is a uniformly hyperbolic set if it is closed and consists of trajectories satisfying the conditions* **(A1)** *and* **(A2)** *with the same constants c and λ.*

It is clear that if $A = M$ then f is an Anosov diffeomorphism. Similarly we can define nonuniformly hyperbolic sets. Another term which is to be found in the literature is *Axiom A*. This was coined by Smale (Smale (1967)), and is similar to, but not quite synonymous with, Anosov.

Definition 5.2.11 (non-wandering) *For $f : M \to M$, a point x is non-wandering if for every neighborhood U (see Definition 3.2.7) of x there is an n such that $f^n(U) \bigcap U \neq \emptyset$. The set of all non-wandering points is the non-wandering set $\Omega(f)$.*

Definition 5.2.12 (Axiom A) *A dynamical system is* Axiom A *if its non-wandering set $\Omega(f)$ is (uniformly completely) hyperbolic and its periodic points are dense in $\Omega(f)$.*

It was thought for a long time that the two criteria for Axiom A were equivalent, but it was shown around 1978 that they are independent (Dankner (1978)).

5.3 Pesin theory

Pesin theory, or the theory of smooth nonuniformly hyperbolic dynamical systems, provides the framework for deducing results connected with the ergodic hierarchy for many systems including linked twist maps. It is based on studying Lyapunov exponents.[3] These have already been mentioned in chapter 3 as a crucial tool in applied dynamical systems giving a measure of the long-term rate of expansion or contraction of tangent vectors. Here we define precisely this quantity, and discuss their existence. Let $f : M \to M$ be a C^2 diffeomorphism (that is, a twice differentiable function with continuous second derivative[4]) on a smooth manifold M with an invariant measure μ.

5.3.1 Lyapunov exponents

The Lyapunov exponent $\chi^{\pm}(z, v)$ at a point z in direction v is given by

$$\lim_{n \to \pm\infty} \frac{1}{n} \log \|Df_z^n v\| = \chi^{\pm}(z, v) \tag{5.21}$$

whenever the limit exists. This caveat is very important. It is clear that when z is a periodic point the limit does indeed exist. However for an arbitrary choice of z (i.e., an arbitrary trajectory), the forward limit $\lim_{n \to \infty}$ may not exist, the backward limit $\lim_{n \to -\infty}$ may not exist, or even if both limits exist they may not be equal. Often, and especially in the applied dynamical systems literature, this issue is ignored and Lyapunov exponents are defined only for forward time, and assumed to exist. Fortunately a celebrated theorem exists which justifies this, provided we have an invariant measure. This is due to Oseledec (1968) and gives the conditions to guarantee that Lyapunov exponents exist almost everywhere and that the forward and backward limits are equal. This is the oft-quoted Multiplicative Ergodic Theorem.

---

[3] These predate Pesin considerably, being named after Aleksandr Mikhailovich Lyapunov (1857–1918), who contributed much to the study of stability of motions, and who first developed the ideas of what are now called *Lyapunov methods*.

[4] Actually $C^{1+\alpha}$, $\alpha > 0$ is sufficient for Pesin's results.

Theorem 5.3.1 (Oseledec (1968)) *Let f be a measure-preserving map f : M →
M such that*

$$\text{(OS)} \int_M \log^+ \|Df_z^{(-1)}\| \mathrm{d}\mu(z) < \infty \tag{5.22}$$

where $\log^+(\cdot) = \max\{\log(\cdot), 0\}$. *Then Lyapunov exponents exist* μ-*almost
everywhere and* $\chi^+(x, v) = \chi^-(x, v) \equiv \chi(x, v)$.

The condition (5.22) (that the integral of the logarithm of the absolute value
of the derivative is finite) is equivalent to requiring that $\log^+ \|Df\|$ be an L^1
function, that is, integrable. (In fact if the space M is compact, this condition
is guaranteed if f has only finitely many discontinuities. Throughout this work
we will be concerned only with compact spaces.)

Lyapunov exponents and hyperbolicity are closely connected. Some of these
connections are plain to see. Suppose $f : M \to M$ is uniformly hyperbolic.
Then choosing any $x \in M$, and any $v \in E_x^s$ we see that from condition **(A2)** in
the definition of uniform hyperbolicity that

$$\frac{1}{n} \log \|Df^n v\| \leq \frac{1}{n} \log(c\lambda^n \|v\|)$$

$$= \frac{1}{n} \log c + \log \lambda + \frac{1}{n} \log \|v\|$$

and so

$$\chi_+(x, v) = \lim_{n \to +\infty} \frac{1}{n} \log \|Df^n v\| \leq \log \lambda < 0$$

(if the limit exists) since $0 < \lambda < 1$.

However, the converse argument does not follow. From Lyapunov exponents
(defined as infinite time limits) being bounded away from zero, one cannot
recover the Anosov conditions **(A2)**, which require uniform expansion and
contraction at each iteration. However, seminal work by Pesin (1977) linked
non-zero Lyapunov exponents with nonuniform hyperbolicity.

Example 5.3.1 *The linear hyperbolic map g of Example 5.2.2 is perhaps the
simplest system for which to compute Lyapunov exponents. The fact that the
Jacobian Dg is the same diagonal matrix at every point on every orbit makes*

the computation trivial. To make this explicit, choosing $v_- = (1,0)^T$ we have:

$$\chi^+(x, v_-) = \lim_{n\to\infty} \frac{1}{n} \log \|Dg^n v_-\|$$

$$= \lim_{n\to\infty} \frac{1}{n} \log \left\| \begin{pmatrix} \lambda & 0 \\ 0 & \mu \end{pmatrix}^n \begin{pmatrix} 1 \\ 0 \end{pmatrix} \right\|$$

$$= \lim_{n\to\infty} \frac{1}{n} \log \left\| \begin{pmatrix} \lambda^n \\ 0 \end{pmatrix} \right\|$$

$$= \lim_{n\to\infty} \frac{1}{n} \log |\lambda^n|$$

$$= \lim_{n\to\infty} \frac{1}{n} \log \lambda^n$$

$$= \lim_{n\to\infty} \log \lambda$$

$$= \log \lambda$$

and the same calculation gives

$$\chi^-(x, v_-) = \log \lambda.$$

Since $\lambda < 1$ this gives a negative Lyapunov exponent, representing the contracting direction. We get a positive Lyapunov exponent corresponding to the expanding direction by taking $v_+ = (0,1)^T$, which gives the similar calculation

$$\chi^+(x, v_+) = \chi^-(x, v_+) = \lim_{n\to\infty} \frac{1}{n} \log \left\| \begin{pmatrix} \lambda & 0 \\ 0 & \mu \end{pmatrix}^n \begin{pmatrix} 0 \\ 1 \end{pmatrix} \right\|$$

$$= \log \mu.$$

Notice that in the area-preserving case $\lambda\mu = 1$ the two Lyapunov exponents sum to zero. Choosing any other initial vector $v = av_+ + bv_-$ will yield the larger Lyapunov exponent (providing $b \neq 0$).

Example 5.3.2 *Consider the toral twist map $t : T^2 \to T^2$, $t(x,y) = (x+y, y)$ (mod 1). This has Jacobian*

$$Dt = \begin{pmatrix} 1 & 1 \\ 0 & 1 \end{pmatrix}, \qquad Dt^n = \begin{pmatrix} 1 & n \\ 0 & 1 \end{pmatrix}.$$

It is straightforward to see that vectors of the form $v^* = (v_1, 0)$ undergo no expansion – thus, $Dt^n v^* = v^*$, while more general vectors $v = (v_1, v_2)$ may be stretched under the shear map, but not exponentially so, hence $\chi^\pm(z, v) = 0$

for all points z and all tangent vectors v. This map forms the basis of a linked twist map.

Example 5.3.3 *Lyapunov exponents for the Cat Map (Example 5.2.4) can be computed almost as simply. Diagonalizing the Jacobian matrix Df produces the two Lyapunov exponents $\chi_1 = \log \lambda_+ > 0$ and $\chi_2 = \log \lambda_- < 0$, where $\lambda_\pm$ are the eigenvalues of the Jacobian computed in Example 5.2.4.*

The existence and nature of the invariant measure is crucial for Oseledec's theorem. To illustrate this we give the following example, which can also be found in Chernov & Markarian (2003):

Example 5.3.4 (Chernov & Markarian (2003)) *Let $f : S^1 \to S^1$ be a map on the circle given by $f(x) = x + 1/3\pi \sin 2\pi x$. The map has fixed points at $x_0 = 0$ and $x_1 = 1/2$. We have the Jacobian $Df = 1 + 2/3 \cos 2\pi x$, so the Lyapunov exponents at the fixed points are easily computed as*

$$\chi(x_0) = \log |Df(x_0)| = \log(5/3),$$

$$\chi(x_1) = \log |Df(x_1)| = \log(1/3).$$

A simple stability analysis on the signs of the Lyapunov exponents shows that points in between x_0 and x_1 (that is, $p \in (0, 1/2)$) tend to x_1 under forward iteration of f and to x_0 under backward iteration (that is, $f^n(p) \to 1/2$ and $f^{-n}(p) \to 0$ as $n \to \infty$). Then for any non-zero vector $v \in TS^1$ we have the limiting behaviour:

$$\lim_{n \to \infty} \frac{1}{n} \log |Df^n v| = \log(1/3),$$

$$\lim_{n \to \infty} \frac{1}{n} \log |Df^n v| = \log(5/3).$$

and so the forward and backward limits are not equal. Putting this into the context of Theorem 5.3.1 we note that Lebesgue measure is not an invariant measure for the map (f is not area-preserving, but dissipative), and so we should not expect Lyapunov exponents to exist Lebesgue-almost everywhere.

5.3.2 Lyapunov exponents and hyperbolicity

Theorem 5.3.2 (Pesin (1977)) *A trajectory $\{f^k x\}$ is nonuniformly (completely) hyperbolic if*

$$\chi(x, v) \neq 0 \quad \text{for } \mathbf{all} \quad v \in T_x M.$$

In other words, if the Lyapunov exponent computed from a given point for any given direction is non-zero, then the trajectory starting from that point is nonuniformly hyperbolic. This can be extended to nonuniform hyperbolicity for a diffeomorphism.

Theorem 5.3.3 (Pesin (1977)) *A diffeomorphism* $f : M \to M$ *is nonuniformly (completely) hyperbolic if for almost every* $x \in M$

$$\chi(x, v) \neq 0 \quad \text{for all} \quad v \in T_x M.$$

Note that we only require non-zero Lyapunov exponents for *almost every* $x \in M$. It may be that there are sets of zero measure with zero Lyapunov exponents. The importance of this theorem is illustrated by the fact that nonuniformly (completely) hyperbolic systems are frequently named *systems with non-zero Lyapunov exponents*. Similar theorems also exist for partial hyperbolicity.

Theorem 5.3.4 (Pesin (1977)) *A trajectory* $\{f^k x\}$ *is nonuniformly partially hyperbolic if*

$$\chi(x, v) \neq 0 \quad \text{for some} \quad v \in T_x M.$$

Theorem 5.3.5 (Pesin (1977)) *A diffeomorphism* $f : M \to M$ *is nonuniformly partially hyperbolic if for almost every* $x \in M$

$$\chi(x, v) \neq 0 \quad \text{for some} \quad v \in T_x M.$$

Note that the fact that partially hyperbolic systems are defined in terms of subspaces which do not span the entire tangent space is made explicit in this formulation.

5.3.3 Stable and unstable manifolds

Pesin theory extends the Stable Manifold Theorem (Theorem 5.2.2) to nonuniformly hyperbolic systems in the usual way – the uniformity of bounds in the uniformly hyperbolic definition is replaced by a function which can vary along the trajectory.

Theorem 5.3.6 (Pesin (1977), Bunimovich *et al.* **(2000))** *Let* $\{f_x^k\}$ *be a nonuniformly hyperbolic trajectory. Then there exists a local stable manifold* $\gamma^s(x)$ *in a neighbourhood* $B(x, r(x))$ *of* x *such that for* $y \in \gamma^s(x)$ *the distance between* $f^k(x)$ *and* $f^k(y)$ *decreases with an exponential rate. That is, for any* $k \in \mathbb{Z}$ *and* $n \geq 0$

$$d(f^{k+n}(x), f^{k+n}(y)) \leq K c(f^k(x), \epsilon) \lambda^n e^{\epsilon n} d(f^k(x), f^k(y)) \qquad (5.23)$$

where d is the distance induced by the Riemannian metric and $K > 0$ is a constant.

Similarly one can show existence of a local unstable manifold. A crucial, but very technical, difference between the local stable manifolds for a uniformly hyperbolic system and a nonuniformly hyperbolic system is in the size of the manifolds. Recall that for the uniformly hyperbolic case $\gamma^s(x)$ was defined[5] in an ϵ-ball of radius $r(x)$ about x, and that $r(x) \geq r' > 0$ for all $x \in M$. In the nonuniform case the situation is different. Here the size of the γ^s are not uniformly bounded away from zero, but satisfy $r(f^k(x)) \geq Ke^{-\epsilon|k|}r(x)$. Thus the size of $\gamma^s(f^k(x))$ may become very small, and indeed may decrease at a rate faster than that given by (5.23). Nevertheless these precise bounds allow many important properties to be formulated for local stable and unstable manifolds for nonuniformly hyperbolic systems (see Bunimovich *et al.* (2000), Pesin (1977) and Barreira & Pesin (2002)).

Example 5.3.5 *The twist map $t : \mathbb{T}^2 \to \mathbb{T}^2$ of Example 5.3.5 has zero Lyapunov exponents, and so the above theorem does not apply. Hence there are no local stable and unstable manifolds for any points in $\mathbb{T}^2$. We shall shortly see the effect this has on the ergodic properties of t.*

5.3.4 Ergodic decomposition

If Lyapunov exponents exist, Pesin's landmark theorem equates the set of points (which we will call Λ) for which Lyapunov exponents are non-zero with the set of points for which f is a nonuniformly hyperbolic diffeomorphism, and guarantees the existence of stable and unstable manifolds. Furthermore, it demonstrates that this set of points has an *ergodic partition*, the definition of which is contained in the statement of the theorem:

Theorem 5.3.7 (Pesin (1977)) *Let*

$$\Lambda = \{x \in M : \chi(x, v) \neq 0 \text{ for every } v \neq 0 \text{ in } T_x M\}$$

and let $\mu(\Lambda) > 0$. Then Λ is either a finite or countably infinite union of disjoint measurable sets $\Lambda_0, \Lambda_1, \ldots$ such that

1. *$\mu(\Lambda_0) = 0$, $\mu(\Lambda_n) > 0$ for $n > 0$,*
2. *$f(\Lambda_n) = \Lambda_n$,*
3. *$f|_{\Lambda_n}$ is ergodic.*

---

[5] We abuse the notation here in order to convey the essence of the idea. In fact $\gamma^s(x)$ is constructed via a map from $B(x, r(x))$ to the unstable subspace. The local stable manifold is then the projection of that map under the exponential map.

Here the notation $f|_{\Lambda_n}$ symbolizes the restriction of the map f to the set Λ_n. We may refer to the sets Λ_n as *ergodic components*. The significance of this theorem is in the fact that we have only finitely or countably many ergodic components in the partition. Any dynamical system can be partitioned into an ergodic partition if an uncountable infinity of components is permitted – trivially decomposing a system into its constituent trajectories creates a partition consisting of uncountably many components of measure zero, and the restriction of a dynamical system to a single trajectory is necessarily ergodic. Moreover, suppose an area-preserving dynamical system has an elliptic fixed point p. Then KAM theory tells us that a positive measure neighbourhood of p contains trajectories which form invariant tori. Even after a perturbation a positive measure set of invariant tori persists. This is another example of a decomposition into an uncountable number of ergodic components. KAM tori, commonly referred to as islands (especially in a fluid mechanics setting), serve as barriers to ergodicity and therefore to mixing. Therefore Theorem 5.3.7 is extremely significant – it guarantees that islands cannot appear in a (completely) nonuniformly hyperbolic system.

Example 5.3.6 *Since the Cat Map has non-zero Lyapunov exponents, we can apply the above theorem to state that it is the union of at most countably many ergodic components. We shall shortly see how to extend this to ergodicity on all of $\mathbb{T}^2$.*

Example 5.3.7 (Pollicott (1983)) *The twist map $t : \mathbb{T}^2 \to \mathbb{T}^2$ of Example 5.3.5 has zero Lyapunov exponents and so Theorem 5.3.7 does not apply. Because there are no stable and unstable manifolds the action of this map is to decompose the torus into an uncountable number of invariant circles $\{(x, y)|x \in [0, 1] \text{ and } y = Y, \text{ fixed}\}$ which are rotated by t through an angle Y. Recall from Example 3.6.1 that the dynamics on each circle is ergodic if Y is irrational, and not ergodic if Y is rational.*

5.3.5 Ergodicity

In applications we are likely to at least want to know how many ergodic components we have in the ergodic decomposition, and more probably would like to show that we have a single ergodic component of full measure. For Anosov systems we can use Theorem 5.3.7 together with the uniform nature of the local stable and unstable manifolds to give a sketch proof of ergodicity. See Barreira & Pesin (2002) for a more rigorous treatment of the argument.

Theorem 5.3.8 (Anosov (1969), Barreira & Pesin (2002)) *Let* $f : M \to$
M be a measure-preserving Anosov diffeomorphism on a connected compact
Riemannian manifold M. Then f is ergodic.

Sketch proof of ergodicity Since f is uniformly hyperbolic, local stable and
unstable manifolds exist and have size $\geq r'$ for each $x \in M$. Assume that each
ergodic component Λ_i is open (see Pesin (1977) or Barreira & Pesin (2002)
for a proof of this assumption). Because the dynamics on Λ_i is ergodic, the
time averages $\phi^+(x)$ of every measurable function ϕ^+ exist and are equal for
almost every $x \in \Lambda_i$ (see Definition 3.6.3). This is true for each i, yet the time
averages might be different on each ergodic component; that is, we may have
$\phi^+(x) \neq \phi^+(y)$ for $x \in \Lambda_i$, $y \in \Lambda_j$, $i \neq j$.
  However, because $\gamma^s(s)$ and $\gamma^u(x)$ have size at least $r' > 0$, we can find
$x_i \in \Lambda_i$ such that $\gamma^s(x_i) \cap \Lambda_j \neq \emptyset$. Recalling Lemma 5.2.1 we see that $\phi^+(x_i) =$
$\phi^+(x_j)$ for all $x_j \in \gamma^s(x) \cap \Lambda_j$, and so time averages are equal on Λ_i and Λ_j.
Since M is connected, any two ergodic components can be joined by a path of
such local stable manifolds (called a *Hopf chain*), and so time averages on each
ergodic component are equal, and so the system is ergodic. □

  Comparing this proof with Section 3.6.1 we see that the uniformity of sizes
of local stable and unstable manifolds gives a way to link together sets (ergodic
components) on which time averages are locally constant. In the nonuniformly
hyperbolic case Pesin (1977) uses topological transitivity (Definition 3.6.2) to
achieve this.

Theorem 5.3.9 (Pesin (1977)) *Let a diffeomorphism f satisfy the conditions*
of Theorem 5.3.7. If f is also topologically transitive then $f|_\Lambda$ is ergodic.

Sketch proof Referring to Theorem 5.3.7, let Λ_i and Λ_j be two distinct ergodic
components such that $\mu(\Lambda_i) > 0$ and $\mu(\Lambda_j) > 0$. Because they are distinct
ergodic components we must have $\mu(f^n(\Lambda_i) \cap \Lambda_j) = 0$ for each $n \in \mathbb{N}$. Now
assume that Λ_i and Λ_j are open sets (see Pesin (1977) or Barreira & Pesin (2002)
for a proof of this assumption). Because f is topologically transitive we have
$f^m(\Lambda_i) \cap \Lambda_j \neq \emptyset$ for some m, and so $\mu(f^m(\Lambda_i) \cap \Lambda_j)) > 0$ (from the openness
of Λ_i and Λ_j). This contradiction implies that the initial definition of Λ_i and Λ_j
as two distinct ergodic components is not valid, and so there can only be one
positive measure component in the ergodic partition. Hence $f|_\Lambda$ is ergodic. □

  Both of the above proofs as we have given them are merely sketch proofs.
We have ignored a large technical issue based on the fact that while $\gamma^s(x)$
and $\gamma^u(x)$ are as smooth as f (and so are differentiable for diffeomorphisms),
they need not depend differentiably on the point x. In order to overcome

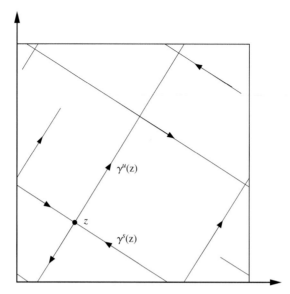

Figure 5.3 Global stable and unstable manifolds for the Arnold Cat Map are straight lines wrapping around the torus, with irrational slopes. Because they fill out a dense set on the torus, and go in different directions, they must intersect.

this difficulty, a property known as absolute continuity of local stable and unstable manifolds must be established. This was done by Anosov (1969) for Anosov diffeomorphisms, and Pesin (1977) for nonuniformly hyperbolic systems. We will not discuss this issue further, as for our purposes we can appeal to theorems such as Theorems 5.3.7 and 5.4.1 directly. For excellent explanations of absolute continuity in these situations, see Barreira & Pesin (2002) or Brin & Stuck (2002).

Example 5.3.8 *The Arnold Cat Map is topologically transitive. A formal proof is given in Theorem 5.1 of Robinson (1998), based roughly on the fact that stable and unstable manifolds of points in $\mathbb{T}^2$ have irrational slopes, and so are dense in $\mathbb{T}^2$ (see Figure 5.3). Therefore by Theorem 5.3.9 the Cat Map is ergodic. (Of course, we have already shown the Cat Map to be an Anosov diffeomorphism, and so Theorem 5.3.8 gives this result automatically.)*

5.3.6 Bernoulli components

Furthermore, Pesin also proves a result about the Bernoulli property, known as the Spectral Decomposition Theorem for nonuniformly hyperbolic diffeomorphisms.

Theorem 5.3.10 (Pesin (1977)) *For each $n \geq 1$ we have*

1. Λ_n *is a disjoint union of sets* Λ_n^m, $m = 1, \ldots, i_n$, *which are cyclically permuted by f (that is, $f(\Lambda_n^m) = \Lambda_n^{m+1}$ for $m = 1, \ldots, i_n - 1$, and $f(\Lambda_n^{i_n}) = \Lambda_n^1$),*
2. $f^{i_n}|_{\Lambda_n^m}$ *is a Bernoulli automorphism for each m.*

So each component in the ergodic decomposition can be repartitioned into components on which some iterate of f restricted to that component is Bernoulli. This result is not of immediate use for us as we are interested in the Bernoulli property of the original map f on the whole domain. Shortly we will see the result which allows us to deduce this, but first we motivate this with a very brief sketch proof of the Bernoulli property for topologically mixing Anosov diffeomorphisms.

Theorem 5.3.11 (Anosov (1969), or see Barreira & Pesin (2002)) *Let $f : M \to M$ be a measure-preserving Anosov diffeomorphism on a connected compact Riemannian manifold, and let f be topologically mixing. Then f has the Bernoulli property.*

Sketch proof Assume the sets Λ_n^m are open sets (again this is proved in Pesin (1977) and Barreira & Pesin (2002)). Then the fact that f is topologically mixing implies that the intersection $f^j(\Lambda_n^m) \cap \Lambda_{n'}^{m'} \neq \emptyset$ for all sufficiently large j, and moreover this intersection is of positive measure. From this one can deduce that the Bernoulli property present on each Λ_n^m extends to the whole domain.

Key point: Pesin theory equates a set of points giving rise to non-zero Lyapunov exponents to nonuniformly hyperbolic behaviour. It also forms a bridge from hyperbolicity theory to the ergodic hierarchy, since it shows that the set of points with non-zero Lyapunov exponents can be partitioned into at most a countable number of components on which the dynamics is ergodic. If the system is topologically transitive then the system itself is ergodic, without the need for a partitioning. Moreover, each of the components can be partitioned further into components on which the Bernoulli property is displayed.

5.4 Smooth maps with singularities

A further complication arises for the class of maps in which we are interested. The derivative Df which we require to compute Lyapunov exponents only exists at points for which f is differentiable. We shall see that certain linked twist maps may contain a set of points (corresponding to the boundaries of the linked annuli) where differentiability fails (a *singularity*). However, such points are rare compared to points at which the linked twist map is differentiable. In such a case we turn to the extension of the Pesin result due to Katok and Strelcyn (Katok *et al.* (1986)) for maps with singularities. This originated from the recognition that 'some important dynamical systems occurring in classical mechanics ... have singularities'. In particular, systems of billiards, which have been much studied by ergodic theorists, often contain singularities.

We must first define precisely what we mean by maps with singularities. Let M be a compact metric space with a metric ρ. We can think of M as being the union of a finite number m of compact Riemannian manifolds $M_1, M_2, \ldots, M_m$ on which a diffeomorphism f will be smooth. The M_i are 'glued together' along a finite number of (C^1) manifolds contained in G, a union of a finite number of (C^1) compact submanifolds. Our singularities will lie along such 'joins'. Let $V = M \backslash G$ be an open dense subset of M (so that V is the original domain with the 'joins' removed). Let μ be a measure on M with $\mu(M) < \infty$ and let f be a μ-invariant, C^2, one-to-one mapping defined on an open set $N \subset V$ into V. Denote $\text{sing}f = M \backslash N$ (this is the set of 'singular' points). Then f is a *smooth map with singularities*. This is a technical description of the idea of removing parts of the domain on which differentiability fails. The Katok–Strelcyn conditions which follow guarantee that the removed parts are not too significant, and we shall see in Chapters 6 and 8 that for the systems of interest to us, smooth maps with singularities which satisfy these conditions are straightforward to define.

5.4.1 Katok–Strelcyn conditions

The work of Katok *et al.* (1986) is based on two key technical conditions. Recalling the definition of an open ϵ-ball $B(\cdot, \epsilon)$ (Definition 3.2.5) we have:

(KS1) There exist constants a, $C_1 > 0$ such that for every $\epsilon > 0$

$$\mu(B(\text{sing}f, \epsilon)) \leq C_1 \epsilon^a, \tag{5.24}$$

and

(KS2) There exist constants b, $C_2 > 0$ such that for every $x \in M \backslash \text{sing}f$

$$\|D^2 f\| \leq C_2 (\text{dist}(x, \text{sing}f))^{-b}. \tag{5.25}$$

Condition **(KS1)** ensures that the 'number' of singularities in our domain is sufficiently small (in fact it implies that $\mu(\text{sing}f) = 0$), and that the measure μ is not concentrated too close to the singularities. Condition **(KS2)** concerns the growth of f, and ensures that the second derivative does not grow too fast near singularities. Note that in Katok *et al.* (1986) three conditions are given. The third is the Oseledec condition ensuring the existence of Lyapunov exponents that we have given as **(OS)** (Equation 5.22).

5.4.2 Ergodicity and the Bernoulli property

The main theorem of Katok *et al.* (1986) pulls together most of the ideas in this chapter for smooth maps with singularities. We introduce the terms Manifold Intersection Property, and Repeated Manifold Intersection Property for properties defined in the following theorem.

Theorem 5.4.1 (Katok *et al.* **(1986))** *Let* $(M, \mathcal{A}, f, \mu)$ *be a measure-preserving dynamical system such that* f *is a smooth map with singularities as defined above.*

(a) Suppose f *satisfies the conditions* **(KS1)**, **(KS2)** *and* **(OS)**. *Then for almost every* $z \in M$ *and for all tangent vectors at* z, *Lyapunov exponents exist, and local stable manifolds* $\gamma^s(z)$ *and local unstable manifolds* $\gamma^u(z)$ *exist for each* z *such that the Lyapunov exponents are negative and positive respectively.*

(b) Suppose in addition all Lyapunov exponents $\neq 0$ *almost everywhere. Then* M *decomposes into a countable family of positive measure,* f-*invariant pairwise disjoint sets* $M = \bigcup_{i=1} \Lambda_i$ *such that* $f|_{\Lambda_i}$ *is ergodic and* $\Lambda_i = \bigcup_{j=1}^{n(i)} \Lambda_i^j$ *where for each* j, $f^{n(i)}|_{\Lambda_i^j}$ *has the Bernoulli property.*

(c) Suppose in addition for almost every $z, z' \in M$ *there exists integers* m, n *such that* $f^m(\gamma^u(z)) \cap f^{-n}(\gamma^s(z')) \neq \emptyset$. *We call this the* **Manifold Intersection Property**. *Then in the decomposition of* M *we have only one set* $\Lambda_i = \Lambda_1$. *This implies* f *is ergodic.*

(d) Suppose in addition for almost every $z, z' \in M$ *and every pair of integers* m, n *sufficiently large,* $f^m(\gamma^u(z)) \cap f^{-n}(\gamma^s(z')) \neq \emptyset$. *We call this the* **Repeated Manifold Intersection Property**. *Then all powers of* f *are ergodic. This implies* f *has the Bernoulli property.*

The proof of this theorem can be found in Katok *et al.* (1986), and is extremely technical. We give the briefest of motivating arguments by showing that the Manifold Intersection Property implies topological transitivity. This corresponds to (c) in Theorem 5.4.1 since we have already shown that topological transitivity implies ergodicity in this situation.

Lemma 5.4.1 *Let the measure-preserving dynamical system* $(M, \mathcal{A}, f, \mu)$ *be such that conditions* **(KS1)**, **(KS2)** *and* **(OS)** *are satisfied, and Lyapunov exponents* $\neq 0$ *almost everywhere. Suppose we also have the Manifold Intersection Property, i.e., for almost every* $z, z' \in M$ *there exists integers* m, n *such that* $f^m(\gamma^u(z)) \cap f^{-n}(\gamma^s(z')) \neq \emptyset$. *Then* f *is topologically transitive.*

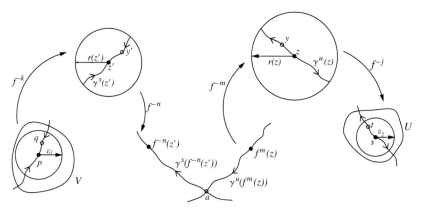

Figure 5.4 Sketch of the construction of a chain from an open set V to an open set U. The filled black circles represent the points p, $z' = f^{-k}(p)$, s and $z = f^j(s)$. The open circles represent the chain itself. This is $t = f^{-j}(y)$, $y = f^{-m}(a)$, $a = f^{-n}(y')$, $y' = f^{-k}(q)$ so that $q = f^i(t)$ with $i = k + n + m + j$.

Sketch proof The general strategy of the proof is to construct a chain between any two arbitrary open sets U and V, in the sense that U contains a point which after some number i of iterations of f ends up in V. Such a chain is illustrated in Figure 5.4 and we use the following facts to allow its construction:

(i) Open sets by definition contain ϵ-balls with positive measure.

(ii) Since the arbitrary set V is open and since $\gamma^s(x)$ exists for almost every $x \in M$, there exists an ϵ-ball $B_1(p, \epsilon_1) \subset V$ about p of radius ϵ_1 such that $\gamma^s(p)$ exists.

(iii) For almost every $z' \in M$, $\gamma^s(z')$ exists and has size $r(z')$; that is, $\gamma^s(z')$ exists in an ϵ-ball $B_2(z', r(z'))$ of radius $r(z')$ (we are not quite rigorous here, for the reasons described in Section 5.3.3). Then $\gamma^s(z')$ has the property that for all $y' \in \gamma^s(z') \subset B_2(z', r(z'))$ there exists an integer k such that $d(f^k(z'), f^k(y')) < \epsilon_1$ (this comes from Equation (5.23)).

(iv) Now from the manifold intersection property, if z' is such that $\gamma^s(z')$ exists, there exists a point $y' \in \gamma^s(z')$ such that $f^{-n}(y') \in f^m(\gamma^u(z))$ for almost every $z \in M$. Call the intersection point a.

(v) Similarly to (i), for any open U there exists a point s such that $\gamma^u(s)$ exists and there exists an ϵ-ball $B_3(s, \epsilon_2)$ about s of radius ϵ_2 contained in U.

(vi) Similarly to (ii), for almost every $z \in M$, there exists an ϵ-ball $B_4(z, r(z))$ of radius $r(z)$ such that $\gamma^u(z)$ exists in $B_4(z, r(z))$.

Now we use the above facts to contruct the chain, referring to Figure 5.4. Given any open set V and a fixed $\epsilon_1 > 0$ there exists a point p and a point

$q \in B_1(p, \epsilon_1) \subset V$ such that $y' = f^{-k}(q) \in \gamma^s(z') = \gamma^s(f^{-k}(p))$ for some k. Similarly, given any open set U and a fixed $\epsilon_2 > 0$ there exists a point s and a point $t \in B_3(s, \epsilon_2) \subset U$ such that $y = f^j(t) \in \gamma^u(z) = \gamma^u(f^j(s))$ for some j. Then by the Manifold Intersection Property, q and t can be chosen so that $f^{-n}(f^{-k}(q)) = f^{-n}(y') = a = f^m(y) = f^m(f^k(t))$. Therefore, setting $i = k + n + m + j$, we have $t = f^{-i}(q)$.

We have shown that for any pair of open sets U and V, there exists an ϵ-ball contained in V of arbitrarily small radius containing a point q such that $f^{-i}(q) = t$, where t is contained in an ϵ-ball of arbitrarily small radius contained in U. Therefore $f^{-i}(V) \cap U \neq \emptyset$, and so $f^i(U) \cap V \neq \emptyset$, proving topological transitivity.

We note here that this is only a sketch proof because, as discussed in Section 5.3.3, we do not have uniform bounds on the sizes $r(z)$ of local stable and unstable manifolds, nor on the rate at which a point $y \in \gamma^s(z)$ approaches z. $\quad\square$

Lemma 5.4.2 *Let the measure-preserving dynamical system* $(M, \mathcal{A}, f, \mu)$ *be such that conditions* **(KS1)**, **(KS2)** *and* **(OS)** *are satisfied, and Lyapunov exponents* $\neq 0$ *almost everywhere. Suppose we also have the Repeated Manifold Intersection Property, so that for almost every* $z, z' \in M$ *and every pair of sufficiently large integers* $m, n, f^m(\gamma^u(z)) \cap f^{-n}(\gamma^s(z')) \neq \emptyset$. *Then* f *is topologically mixing.*

Sketch proof The repeated manifold intersection property ensures that we can construct a family of chains in the manner of the previous proof such that for any open set V, there is a point $q \in V$ such that $f^{-i}(q) = t$, where t is contained in an arbitrary open set U, where $i = k + n + m + j$, for all sufficiently large n and m. Therefore $f^i(U) \cap V \neq \emptyset$ for all sufficiently large i, showing topological mixing. $\quad\square$

The result of Theorem 5.4.1, in the words of the authors of Katok *et al.* (1986), is to 'generalize Pesin's results to a broad class of dynamical systems with singularities and at the same time to fill gaps and correct errors in Pesin's proof of the absolute continuity of families of invariant manifolds'. The topological conditions in parts (c) and (d) of the theorem are closely linked to the ideas of topological transitivity and topological mixing, as the two previous lemmas demonstrate.

Key point: Katok–Strelcyn theory allows the extension of Pesin theory for systems with singularities, such as non-smooth linked twist maps, and adds a condition to verify the Bernoulli property on the entire domain.

Example 5.4.1 *The Arnold Cat Map is a smooth map, so results for smooth maps with singularities also apply to it. It satisfies all the conditions in Theorem 5.4.1, and so has the Bernoulli property. Of course, we do not need to appeal to machinery as heavy as the Katok–Strelcyn theorem to show this,[6] but this is an important building block to understand similar results for linked twist maps.*

5.5 Methods for determining hyperbolicity

5.5.1 Invariant cones

To apply Pesin's theorem, we must verify that Lyapunov exponents exist, and then show that they are not equal to zero. Note that if we wish the ergodic partition to cover the whole of our original domain we must have non-zero Lyapunov exponents almost everywhere (that is, at least on a set of full measure). For this we turn to the Anosov conditions for hyperbolicity above in Equations (5.3) and (5.4). If these inequalities hold then Lyapunov exponents (if they exist) have absolute value at least $\log |\lambda|$. Demonstrating that these inequalities hold equates to showing that the diffeomorphism is uniformly hyperbolic. As we have seen, in simple systems such as the Cat Map, we might be able to produce the desired splitting of tangent space directly, by knowing the exact directions of stable and unstable manifolds. In general however, this is more complicated. We do not usually know where E^s and E^u lie. To overcome this we employ a technique known variously as the method of invariant cones, invariant sectors, or Alekseev's method. The technique was devised in Alekseev (1969), and also appears in Sinai (1970), Moser (1973), and was adapted in Wojtkowski (1985). Discussions of the method can be found in, for example, Katok & Hasselblatt (1995), Chernov & Markarian (2003), Brin & Stuck (2002) and Robinson (1998).

The general idea is to find a cone, or sector, in the tangent space which is mapped (strictly) into itself under forward iteration of the map (and its Jacobian), such that vectors within that cone are expanded. Such a cone is called an *unstable cone*. This idea was effectively discussed in Section 2.2.1, and there is a direct analogy here with Figure 2.6. There we saw how a sector produced by a pair of initial directions was contracted under forward iteration, and homed in on the direction of the unstable manifold. Recall also how the 'length' of the iterated

---

[6] For example, the method of *Markov partitions* can be used to extract the symbolic dynamics directly. See, for example, Adler (1998).

vectors grew. This method will also work even if the unstable manifolds do not simply lie in the same direction at every point, providing our initial choice of cone encompasses all possible directions in which the unstable manifolds could lie. Because for hyperbolicity we require linear contracting behaviour as well, we must also find a stable cone which maps (strictly) into itself under backward iteration, containing vectors which are expanded under those backward iterations.

We will give the construction in more detail in the case of a two-dimensional diffeomorphism $f : \mathbb{R}^2 \to \mathbb{R}^2$. The generalization to maps on the torus and to higher dimensional maps is relatively straightforward. First we define a cone in general. For a point $x \in \mathbb{R}^2$ and tangent vector $v \in T\mathbb{R}^2$, let $v = v_1 + v_2$, with $v_1 \in L$, a subspace of $T\mathbb{R}^2$, and $v_2 \in L^\perp$, an orthogonal subspace of $T\mathbb{R}^2$ (in two dimensions this simply means we write any vector v as a linear combination of two chosen basis vectors).

Definition 5.5.1 (cone) *(Katok & Hasselblatt (1995), Chernov & Markarian (2003)) A cone of size γ at $x \in \mathbb{R}^2$ is defined to be the set*

$$C_\gamma(x) = \{v_1 + v_2 \in T_x\mathbb{R}^2 \text{ such that } |v_1| \leq \gamma |v_2|\}.$$

Drawing such a cone in tangent space reveals why they are sometimes called sectors. In higher dimensions the cones may have a more general geometry.

Now we define a cone for almost all $x \in \mathbb{R}^2$. Gathering together all such cones we form a cone field (also called a sector bundle) C:

Definition 5.5.2 (cone field) *The collection $\{C_\gamma(x)\}$ given by*

$$\{C_\gamma(x)\} = \bigcup_{a.e.x\in\mathbb{R}^2} C_\gamma(x)$$

is a cone field *on* $\mathbb{R}^2$.

Definition 5.5.3 *A cone field is* invariant *under f if, for almost every $x \in \mathbb{R}^2$,*

$$DfC_\gamma(x) \subseteq C_\gamma(f(x)).$$

A cone field is strictly invariant *under f if, for almost every $x \in \mathbb{R}^2$,*

$$DfC_\gamma(x) \subset C_\gamma(f(x)).$$

A cone field is eventually strictly invariant *under f if, for almost every $x \in \mathbb{R}^2$, there exists $n \in \mathbb{N}$ (which may depend on x) such that*

$$D_x f^n C_\gamma(x) \subset C_\gamma(f^n(x)).$$

We will require two cones, one for the expanding direction and one for the contracting direction. Thus we choose $L = E^s$, $L^\perp = E^u$, and let $v_1 = v_s$,

$v_2 = v_u$ so that $v = v^s + v^u$ has stable and unstable components. Now define the pair of cones

$$C_\gamma^s(x) = \{v^s + v^u \in T_x\mathbb{R}^2 \text{ such that } |v^u| \leq \gamma|v^s|\},$$

$$C_\gamma^u(x) = \{v^s + v^u \in T_x\mathbb{R}^2 \text{ such that } |v^s| \leq \gamma|v^u|\}.$$

We observe that frequently such a pair of cones are referred to as *horizontal* and *vertical* cones (see for example Katok & Hasselblatt (1995)). This is chiefly to stress the transversal nature of the construction, but as we shall see in later chapters, cones need not be centered around the horizontal and vertical axes of tangent space. They key point is that they correspond to stable and unstable directions.

Definition 5.5.4 *A cone field* $\{C_\gamma^s\}$ *is* contracting, *if for almost every* $x \in \mathbb{R}^2$ *and for all* $v \in C_\gamma^s(x)$,

$$|Df(v)| < |v|.$$

Similarly, a cone field $\{C_\gamma^u\}$ *is* expanding, *if for almost every* $x \in \mathbb{R}^2$ *and for all* $v \in C_\gamma^u(x)$,

$$|Df(v)| > |v|.$$

A cone field $\{C_\gamma^s\}$ *is* interior-contracting, *if for almost every* $x \in \mathbb{R}^2$ *and for all* $v \in int(C_\gamma^s(x))$,

$$|Df(v)| < |v|.$$

Similarly, a cone field $\{C_\gamma^u\}$ *is* interior-expanding, *if for almost every* $x \in \mathbb{R}^2$ *and for all* $v \in int(C_\gamma^u(x))$,

$$|Df(v)| > |v|.$$

Proposition 5.5.1 (see e.g., Brin & Stuck (2002)) *Let* $f : M \to M$ *be a diffeomorphism of a compact space. If there exist cone fields* $\{C_\gamma^s\}$ *and* $\{C_\gamma^u\}$ *such that*

1. $\{C_\gamma^s\}$ *is strictly invariant under* f^{-1}
2. $\{C_\gamma^u\}$ *is strictly invariant under* f
3. $\{C_\gamma^s\}$ *is contracting*
4. $\{C_\gamma^u\}$ *is expanding*

then f *is uniformly hyperbolic.*

Under these strict conditions for invariance and growth, compactness guarantees that there exists a constant $0 < \lambda < 1$ such that the Anosov conditions are satisfied. In this case the splitting of tangent space is given by

$$E^s(x) = \bigcap_{n \geq 0} Df^{-n}_{f^n(x)} C^s_\gamma (f^n(x)),$$

$$E^u(x) = \bigcap_{n \geq 0} Df^n_{f^{-n}(x)} C^u_\gamma (f^{-n}(x)).$$

We can relax this classical theorem in one of two ways to produce nonuniform hyperbolicity. The requirement of strict invariance under every iterate could be relaxed to give:

Theorem 5.5.1 (Alekseev (1969)) *Let $f : M \to M$ be a diffeomorphism of a compact space. If there exist cone fields $\{C^s_\gamma\}$ and $\{C^u_\gamma\}$ such that*

1. $\{C^s_\gamma\}$ *is invariant under f^{-1}*
2. $\{C^u_\gamma\}$ *is invariant under f*
3. $\{C^s_\gamma\}$ *is contracting*
4. $\{C^u_\gamma\}$ *is expanding*

then f has non-zero Lyapunov exponents almost everywhere.

Alternatively we could relax the requirement of needing contraction (resp. expansion) for *all* vectors in C^s_γ (resp. C^u_γ) to give:

Theorem 5.5.2 (Alekseev (1969)) *Let $f : M \to M$ be a diffeomorphism of a compact space. If there exist cone fields $\{C^s_\gamma\}$ and $\{C^u_\gamma\}$ such that*

1. $\{C^s_\gamma\}$ *is invariant and eventually strictly invariant under f^{-1}*
2. $\{C^u_\gamma\}$ *is invariant and eventually strictly invariant under f*
3. $\{C^s_\gamma\}$ *is interior-contracting*
4. $\{C^u_\gamma\}$ *is interior-expanding*

then f has non-zero Lyapunov exponents almost everywhere.

The growth conditions for tangent vectors v in Definition 5.5.4 are not quite the same as the Anosov conditions (5.3) and (5.4), since the Anosov conditions require the growth of tangent vectors to have a uniform bound λ strictly greater than one. However the fact that v lies in an invariant and eventually strictly invariant cone allows the Anosov conditions to be deduced (heuristically, if we wait long enough the growth will eventually be great enough).

Key point: The cone field argument is a crucial tool for demonstrating that Lyapunov exponents are non-zero.

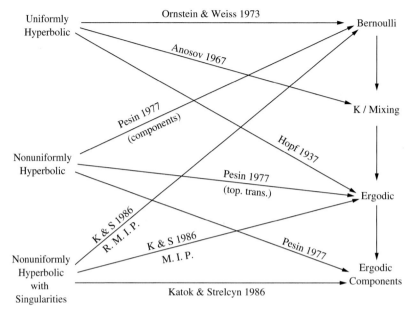

Figure 5.5 Summary of results connecting hyperbolic systems with the ergodic hierarchy.

5.6 Summary

We summarize the development of the links between the theory of hyperbolic dynamical systems and the ergodic hierarchy in Figure 5.5. On the left in order of decreasing strictness are the notions of uniform hyperbolicity, nonuniform hyperbolicity and nonuniformly hyperbolic systems with singularities. On the right in order of decreasing complexity is the ergodic hierarchy, in which the Bernoulli property implies the K-property and mixing, which in turn imply ergodicity, which is a special case of a partition into ergodic components. In this chapter we have given some details of uniformly and nonuniformly hyperbolic systems, and in particular discussed Pesin theory, which guarantees an ergodic decomposition for nonuniformly hyperbolic systems, ergodicity for systems which also possess topological transitivity, and a decomposition into Bernoulli components. We have also given the results of Katok *et al.* (1986). These extend the results of Pesin to smooth maps with singularities, and give the Manifold Intersection Property (M. I. P.) as a condition for ergodicity, and the Repeated Manifold Intersection Property (R. M. I. P.) as a condition for the Bernoulli property.

6

The ergodic partition for toral linked twist maps

This chapter discusses the application of Pesin theory to linked twist maps. Drawing on three key papers from the ergodic theory literature we give the proof that linked twist maps can be decomposed into at most a countable number of ergodic components.

6.1 Introduction

In Chapter 4 we gave Devaney's construction of a horseshoe for a linked twist map on the plane. The existence of the horseshoe and the accompanying subshift of finite type implies that the linked twist map contains a certain amount of complexity. However, topological features such as horseshoes may not be of interest from a statistical, observable, or measure-theoretic point of view, as they occur on invariant sets of measure zero. The subshift of finite type occurs on just such an invariant set of measure zero and is therefore arguably not of significant statistical interest. Nevertheless it is possible that similar behaviour is shared by points in the vicinity of the horseshoe, meaning that complex behaviour is present in a significant (that is, positive measure) domain. Easton (1978) conjectures that this may indeed be the case, and that in fact linked twist maps may be ergodic.

Three papers provide the framework for applying the results of Pesin (1977) connecting hyperbolicity and ergodicity. In this and the following two chapters we draw heavily on each of Burton & Easton (1980), Wojtkowski (1980) and Przytycki (1983). After defining in detail different forms of toral linked twist map, we begin with the work of Burton & Easton (1980), who consider the simplest case – smooth co-rotating linked twist maps on a two-torus – which has the advantages of allowing each twist map to be expressed in the same coordinate system, and also allowing the system to be differentiable at all points. In

Chapter 8 we see how the difficulties of the necessary coordinate changes in the planar case can be overcome by appealing to Wojtkowski (1980). The result proved in Burton & Easton (1980) is that the domain of smooth co-rotating toral linked twist maps is the union of (possibly countably many) ergodic components (an *ergodic partition*). Of course, this is still some way short of the goal of showing that linked twist maps have the Bernoulli property on all of their domain. In the next chapter we discuss the geometrical arguments needed to deduce a single ergodic component, and the extension to mixing and the Bernoulli property, using the argument in Przytycki (1983).

Just as in the Arnold Cat Map in Chapter 5 the key ingredients of the toral linked twist maps are a pair of transverse directions, one in which regions of phase space are expanded, or stretched, and one in which regions are contracted. In the Cat Map, this expansion and contraction occurs at every iterate of the map, while for linked twist maps only iterates landing in the appropriate annulus result in expansion or contraction. Because of this fact, toral linked twist maps can be viewed as a nonuniformly hyperbolic generalization of the Cat Map. The basic strategy behind the proofs in this chapter is to show that toral linked twist maps have non-zero Lyapunov exponents. This is more straightforward for the co-rotating, co-twisting case, as here both twists result in expansion in a common direction. By contrast, in the counter-rotating, counter-twisting case, the vertical twist may contract the expansion achieved by the horizontal twist, and so we require a condition on the strength of the twists to ensure the necessary result.

6.2 Toral linked twist maps

We define here a general form of the toral linked twist map, similar to that found in Burton & Easton (1980), Wojtkowski (1980) and Przytycki (1983). We have already discussed some of the definitions here in Chapter 2 when putting linked twist maps into the context of mixers, but we give the definitions again, with some more detail, for ease and clarity.

Consider the two-dimensional torus $\mathbb{T}^2$ with coordinates (x, y) (mod 1). Define two overlapping annuli P, Q by

$$P = \{(x,y) : y_0 \leq y \leq y_1\}$$
$$Q = \{(x,y) : x_0 \leq x \leq x_1\}.$$

Denote the union of the annuli $R = P \cup Q$ and the intersection $S = P \cap Q$. See Figure 6.1. (For reference, Burton & Easton (1980) take $x_0 = y_0 = 0$,

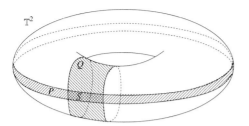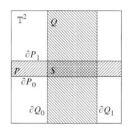

Figure 6.1 The 2-torus $\mathbb{T}^2$ with annuli P and Q. In the square on the right, the upper and lower edges are identified and the left and right edges are identified to create the torus on the left.

$x_1 = y_1 = 1/2$ so R is an L-shaped region covering 3/4 of the torus, while Wojtkowski (1980) takes $x_0 = y_0 = 0$, x_1, y_1 arbitrary, so R is an L-shaped region of arbitrary size. Przytycki (1983) defines his annuli as we have done.) We will frequently need to refer to the boundaries of the annuli P and Q, and so we give the following definitions.

Definition 6.2.1 ($\partial P_0, \partial P_1, \partial Q_0, \partial Q_1, \partial P, \partial Q$) *We denote the lower boundary of P (that is, the line $\{(x,y) : x \in [0,1], y = y_0\}$) by ∂P_0. Similarly we denote the upper boundary of P by $\partial P_1 = \{(x,y) : x \in [0,1], y = y_1\}$, the left boundary of Q by $\partial Q_0 = \{(x,y) : x = x_0, y \in [0,1]\}$, and the right boundary of Q by $\partial Q_1 = \{(x,y) : x = x_1, y \in [0,1]\}$. Finally we denote the unions $\partial P = \partial P_0 \cup \partial P_1$ and $\partial Q = \partial Q_0 \cup \partial Q_1$.*

It is usual and natural to unwrap the torus $\mathbb{T}^2$ and represent it as the unit square, as in Figure 6.1. The top and bottom edges of the square are identified, as are the left and right edges. The annuli P and Q then become vertical and horizontal strips in the square.

6.2.1 Twist maps on the torus

A twist map is defined by assigning to each annulus a shear. In particular we define the function

$$F : R \to R$$

$$F = F(x, y; f) = \begin{cases} (x + f(y), y) & \text{if } (x,y) \in P \\ (x, y) & \text{if } (x,y) \in R \backslash P, \end{cases}$$

where $f : [y_0, y_1] \to \mathbb{R}$ is a real-valued function, called a *twist function*, such that $f(y_0) = 0$ and $f(y_1) = k$, for some integer k. So if F acts on a point (x, y) in R (the union of P and Q) but not in P, it leaves that point unchanged (in other

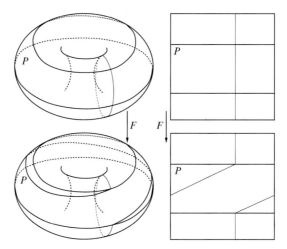

Figure 6.2 Illustration of the action of an F twist in both representations of the torus $\mathbb{T}^2$. In the upper row the boundaries ∂P of P are shown in bold. The thin line is a line of constant x which will act as a line of initial conditions which are transformed under one iterate of F into the thin line in the lower diagrams. Initial conditions outside P remain where they begin while a line of initial conditions within P gets wrapped around the torus. Here we have chosen $k = 1$ so the wrapping number is 1.

words, F is the *identity* map, Id, on $R\backslash P$). If F is applied to a point (x, y) in P, the y-coordinate is left unchanged, but the x-coordinate is altered by an amount dependent on the value of y. We insist that k must be an integer in order that the two components of F should 'join up' at the boundary of P – that is, F should be *continuous* on ∂P. We illustrate this in Figure 6.2, which shows a sketch of the image of a line of constant x under iteration of the map F. Points outside P do not move, while the section of the original line inside P is wrapped k times around the torus (in this figure we have taken $k = 1$). Since k is an integer, the image is also a continuous line. (For case of illustration we have drawn f as a linear function, but this need not be the case. We will discuss this further in the following sections).

We have an exactly analogous twist map assigned to Q:

$$G : R \to R$$

$$G = G(x, y; g) = \begin{cases} (x, y + g(x)) & \text{if } (x, y) \in Q \\ (x, y) & \text{if } (x, y) \in R\backslash Q, \end{cases}$$

where $g : [x_0, x_1] \to \mathbb{R}$ is a real-valued function such that $g(x_0) = 0$ and $g(x_1) = l$, for some integer l. Again $G = Id$ outside the annulus Q. Applying the map G to a line of constant y results in an image which wraps around the

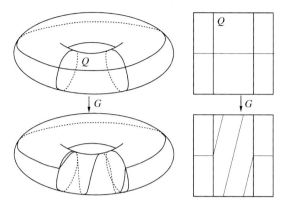

Figure 6.3 Illustration corresponding to Figure 6.2 showing the action of a twist G. The bold lines depict the annulus Q. Here we have chosen $l = 2$ and so the section of the thin line of initial conditions within Q are wrapped around the torus twice.

torus l times before joining up with line segments left unchanged. Figure 6.3 shows this for $l = 2$. For obvious reasons we refer to k and l as the *wrapping number* of the twist.[1] If a twist has wrapping number 1 or 2, we may also refer to it as a single-twist or double-twist respectively, again for obvious reasons. Note that in general k and l may be positive or negative integers and so twists may wrap around the torus in either direction. The choice of sign of kl makes a crucial difference to the ensuing results and methods of proof in the chapter.

Having ensured the continuity of F and G we turn to further smoothness properties. Since the identity map is smooth, the map F will be endowed with the same smoothness properties as the function f if and only if these smoothness properties also hold on ∂P.

There is much to be said about restrictions on the form of the functions f and g. For now we will assume that the functions f and g are C^2 – that is, twice differentiable with continuous second derivatives (this assumption is for technical reasons later on). Further we assume that we have

$$\left.\frac{\mathrm{d}f}{\mathrm{d}y}\right|_y \neq 0 \quad \left.\frac{\mathrm{d}g}{\mathrm{d}x}\right|_x \neq 0$$

for each $y_0 < y < y_1$ and each $x_0 < x < x_1$. This condition, that the derivatives of f and g do not vanish, ensures that we have monotonic increasing or decreasing twists. We now define the *strength* of a twist by considering the

---

[1] Note that this is similar to the concept of a *winding number* or *rotation number* of a circle map (see for example Katok & Hasselblatt (1995)).

shallowest slopes of f and g. Since the twist functions $f(y)$ and $g(x)$ will always be functions of y and x respectively, we will write their derivatives as $f'(y)$ and $g'(x)$.

Definition 6.2.2 (strength of a twist) *The* strength α *of a twist map F on a horizontal annulus P is given by:*

$$\text{if } k > 0 \quad \alpha = \inf \left\{ f'(y) : y_0 < y < y_1 \right\}$$
$$\text{if } k < 0 \quad \alpha = \sup \left\{ f'(y) : y_0 < y < y_1 \right\},$$

and similarly, the strength β *of a twist map G on a vertical annulus Q is given by*

$$\text{if } l > 0 \quad \beta = \inf \left\{ g'(x) : x_0 < x < x_1 \right\}$$
$$\text{if } l < 0 \quad \beta = \sup \left\{ g'(x) : x_0 < x < x_1 \right\}.$$

Both the properties defined above – the wrapping number and the strength of the twist – will be important factors in the behaviour of a linked twist map. Thus we name, as in Przytycki (1983), such a twist map F a (k, α)-twist, and such a twist map G a (l, β)-twist.

6.2.2 Linking the twist maps

Finally the toral linked twist map H is defined by composing F and G:

$$H : R \to R$$
$$H = H(x, y; f, g) = G \circ F.$$

The action of H is illustrated in Figure 6.4. Referring to this figure, it is worth noting explicitly the role that the identity mapping in the definitions of F and G plays as we iterate under H. Suppose an initial point (x, y) is in P. When the map F is applied (the first component of the linked twist map H), the image remains in P. If this image $F(x, y)$ falls into the overlap S then the second component of H (the twist map G) moves our point in a transverse direction, giving the image $H(x, y)$. In Figure 6.4 this is shown in black. Note now what would happen if we applied H again. Our new initial point $H(x, y)$ is in $R \backslash P$, and so the first component of H is simply the identity map, and this would be followed by an iteration of G. Similarly suppose now that we chose an initial condition (x', y') (shown in grey in Figure 6.4) in P which after an iterate of F does not lie in S. Then the contribution from G is the identity map and $F(x', y') = H(x', y')$. This illustrates the two key elements of a linked twist map. First, during iteration a point can be moved in one of two different (transverse) directions (in fact here the directions lie at right angles but this is simply an artefact of the simplicity

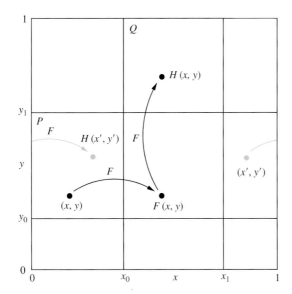

Figure 6.4 The action of the toral linked twist map H. An initial condition (x, y) (shown in black) for which $F(x, y) \in S$ is transformed by the twists F and G in transverse directions. Frequently however, a component of H is the identity map, for example the next iterate of $H(x, y)$, or an initial condition (x', y') (shown in grey) for which $F(x', y') \notin S$. Then $F(x', y') = H(x', y')$ since the contribution from G is the identity.

of the toral framework. In Chapter 8 on planar linked twist maps we will see that this is not necessary, as long as the directions are sufficiently transverse). Second, in between moments when a point may be moved under the action of either F or G, we may have times when the action of F or G is to leave a point stationary. As we shall see, this is the crucial factor in the non-uniformity of a linked twist map's hyperbolicity.

The likelihood of encountering an identity transformation for a given iterate is, roughly speaking, a function of the size of the annuli P and Q. It is clear that in the case $x_0 = y_0 = 0$, $x_1 = y_1 = 1$ we have $P = Q = S$ and so at every iteration of H we have a twist from F and one from G and the identity is never applied. Every iterate of H lands in S. This is a simpler system than a toral linked twist map, called a toral automorphism. For the correct choice of twist functions this is a uniformly hyperbolic system. In particular, the Arnold Cat Map can be regarded as a linked twist map.

Example 6.2.1 (Arnold Cat Map as a linked twist map) *If we set $x_0 = y_0 = 0$ and $x_1 = y_1 = 1$, then the annuli and their union and intersection all coincide*

*on the torus: $P = Q = R = S = \mathbb{T}^2$. Setting the twist functions as $f(y) = y$
and $g(x) = x$, we have $F(x, y) = (x + y, y)$ and $G(x, y) = (x, y + x)$ for all
$(x, y) \in T^2$. Then $H(x, y) = G \circ F = (x + y, x + 2y)$. This is the Arnold Cat
Map, which can be viewed as the composition of two shear maps. Although
it is possible to prove the Bernoulli property for the Arnold Cat Map directly
(using Markov partitions), the fact that it can be cast as a form of linked twist
map means that we can verify this property by proving the result for the linked
twist maps.*

Key point: If we link together maps on annuli which are the width of the whole
torus, we get the simpler situation of a toral automorphism. Choosing the correct
twist functions then gives a uniformly hyperbolic system as the map is subject to
the same expansion and contraction at every iteration. In particular the Arnold Cat
Map is an example of such a system.

6.2.3 First return maps

In the case $P \neq S$ or $Q \neq S$ we will need to handle the situation in which some
iterates of F, G and H land in S and others do not. To this end we construct *first
return maps*.

Definition 6.2.3 (F_S) *The first return map $F_S : S \to S$ is given by*

$$F_S(z) = F^m(z)$$

*for each $z \in S$, where $m = m(z)$ is such that $F^m(z) \in S$ and $F^i(z) \notin S$ for
$1 \leq i < m$.*

Note that the integer m in the above definition is dependent on z. There may be
some points $z \in S$ for which $F^m(z) \notin S$ for all $m > 0$. However, since F is a
rigid rotation on each circle $y = $ constant, the set of such z is a subset of those
points $z = (x, y)$ for which $f(y)$ is rational, and so has measure zero.[2] We define
similarly the first return map for G and H.

Definition 6.2.4 (G_S) *The first return map $G_S : S \to S$ is given by*

$$G_S(z) = G^{m'}(z)$$

*for each $z \in S$, where $m' = m'(z)$ is such that $G^{m'}(z) \in S$ and $G^i(z) \notin S$ for
$1 \leq i < m'$.*

---

[2] Due to the fact that a one-to-one function takes a countable set of points onto a countable set
of points.

Definition 6.2.5 (H_S) *The first return map $H_S : S \to S$ is given by*

$$H_S(z) = H^{m''}(z)$$

for each $z \in S$, where $m'' = m''(z)$ is such that $H^{m''}(z) \in S$ and $H^i(z) \notin S$ for $1 \leq i < m''$.

The reader will find it straightforward to see that

$$H_S = G_S \circ F_S.$$

In the following we frequently drop the composition symbol '$\circ$' for ease of notation. It is worth noting explicitly the relationship between H and H_S. Applications of H form a chain of applications of F and G. If $z \in P$ and $F(z) \in S$ then $H \equiv GF$ (by convention composition of maps takes place on the left). If $z \in P$ and $F(z) \notin S$ then $H \equiv F$ (as $G \equiv \text{Id}$ outside Q). Similarly, if $z \in Q \backslash P$ then $H \equiv G$ (as $F \equiv \text{Id}$ outside P). Applications of H_S also form a chain of F and G, but now H_S is equivalent to a chain of Fs (until we hit S) followed by a chain of Gs (until we hit S). Both H and H_S represent the same long-term dynamics, as can be seen in the following sample chain of maps, (remembering the evolution of composition of maps reads right to left, and noting that here we do not write any application of the identity map Id):

$$
\overbrace{G}^{H} \; \overbrace{G}^{H} \; \overbrace{G}^{H} \; \overbrace{GF}^{H} \; \overbrace{GF}^{H} \; \overbrace{F}^{H} \; \overbrace{G}^{H} \; \overbrace{G}^{H} \; \overbrace{GF}^{H} \; \overbrace{G}^{H} \; \overbrace{GF}^{H} \; \overbrace{F}^{H} \; \overbrace{F}^{H}
$$

$$
\underbrace{}_{H_S} \; \underbrace{}_{H_S} \; \underbrace{}_{H_S} \; \underbrace{}_{H_S}
$$

More precisely, take an initial $z \in S$. Let $F_S(z) = F^{m_0}(z)$ (as defined above), and let $G_S(F_S(z)) = G^{m'_0}(F_S(z))$, so $H_S(z) = G_S \circ F_S = G^{m'_0}(F^{m_0}(z)) = H^{m''_0}(z)$. Note that $m''_0 = m_0 + m'_0 - 1$. Denoting the orbit $z_i = H^i_S(z)$, we see that $F_S(z_i) = F^{m_i}(z_i)$, and similarly for G and m'_i, so that $H^n_S = H^j$, where $j = \sum_{i=0}^{n-1}(m_i + m'_i - 1)$. Of course, each of the m_i, m'_i depends on the initial z.

We note here that Wojtkowski (1980) constructs sequences m_i and m'_i for each z, which count the number of iterates of F and G between visits to S for an trajectory of H. These allow one to calculate explicitly the direction of stable and unstable manifolds of z using continued fractions (see Chapter 7 for more details).

> **Key point:** When constructing a return map to the intersection region, the number of iterations of the original map that is needed to return a point to the intersection is dependent on the point itself. The fact that different points take different amounts of time to return is the feature that makes the hyperbolicity of a linked twist map nonuniform.

6.2.4 Co-rotating toral linked twist maps

For a co-rotating linked twist map we require k and l to be of the same sign. We assume throughout without loss of generality that for the co-rotating case both k and l are positive. (If both are negative then either the transformation $(x, y) \rightarrow (-x, -y)$, or reversing the direction of time, will recover our choice of polarity.) Recall that on the torus co-rotating is the same as co-twisting.

6.2.5 Counter-rotating toral linked twist maps

When k and l are different signs the twists go in different directions. In all of the following we assume without loss of generality that, for the counter-rotating case, $k > 0$ and $l < 0$. The symmetry of the standard torus may suggest that a simple transformation ought to render the co-rotating and counter-rotating cases identical, but this is not the case, for the reasons described in Section 2.2.1, and developed in more detail in this chapter. Recall that on the torus counter-rotating is the same as counter-twisting.

6.2.6 Smooth twists

To make the linked twist map smooth (for our purposes at least twice continuously differentiable, except possibly at the boundaries of P and Q, where we require only continuous differentiability) we require the following conditions on f and g:

$$f'(y_0) = f'(y_1) = g'(x_0) = g'(x_1) = 0.$$

This ensures that on ∂P and ∂Q (the boundaries of P and Q respectively), the value of the derivatives of f and g equal the value of the derivative of the identity. Because we insist that f and g are monotonic, we also have (on the following *open* intervals for the case $k > 0, l > 0$):

$$f'(y) > 0 \text{ for } y \in (y_0, y_1),$$
$$g'(x) > 0 \text{ for } x \in (x_0, x_1).$$

Then the linked twist map is differentiable everywhere, including on the boundaries of P and Q. These are just the conditions on the twists assumed by Burton & Easton (1980). An example of a smooth twist can be given by the polynomial $f(y) = ay^3 + by^2 + cy + d$, where

$$a = \frac{-2k}{-y_0^3 - 3y_0y_1^2 + y_1^3 + 3y_1y_0^2},$$

$$b = \frac{3k(y_1 + y_0)}{-y_0^3 - 3y_0y_1^2 + y_1^3 + 3y_1y_0^2},$$

$$c = \frac{-6ky_0y_1}{-y_0^3 - 3y_0y_1^2 + y_1^3 + 3y_1y_0^2},$$

$$d = \frac{ky_0^2(-y_0 + 3y_1)}{-y_0^3 - 3y_0y_1^2 + y_1^3 + 3y_1y_0^2}.$$

These coefficients guarantee that $f(y_0) = 0, f(y_1) = k$, and $f'(y_0) = f'(y_1) = 0$. We have corresponding cofficients for $g(x) = a'x^3 + b'x^2 + c'x + d'$. We show this smooth twist function in Figure 6.5(b) for $k = 1$. Note that although we have $f'(y) > 0$ for $y \in (y_0, y_1)$ and similarly for g, we have $\alpha = \beta = 0$. This is because the derivatives tend to zero on ∂P and ∂Q. However on an interval $[y_0', y_1']$ where $y_0' > y_0$ and $y_1' < y_1$ the infimum of $f'(y)$ is bounded away from zero (assuming $k > 0$), and similarly for g. This is the basis of the technique used in Section 6.3.1 to prove our desired result.

6.2.7 Non-smooth twists

If the requirement that the derivatives of f and g should vanish at the boundaries of P and Q is released, then our linked twist map is no longer differentiable at points lying on the boundaries of P and Q. Such points are called *singularities* of the map, and the method of dealing with them is discussed in Section 5.3. Since we assume throughout that section that $f'(y) \neq 0$ for $y \in [y_0, y_1]$, and $g'(x) \neq 0$ for $x \in [x_0, x_1]$, these conditions ensure that we can have $\alpha > 0$, and $\beta > 0$ in the co-rotating case, and $\beta < 0$ in the counter-rotating case.

6.2.8 Linear twists

The simplest non-smooth forms for f and g are linear functions, giving the linear twist maps found in Wojtkowski (1980). Specifically these are

$$f(y) = \frac{k}{y_1 - y_0}(y - y_0),$$

$$g(x) = \frac{l}{x_1 - x_0}(x - x_0),$$

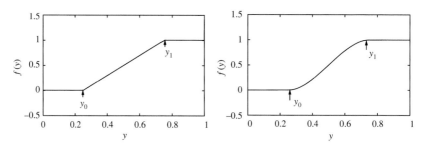

Figure 6.5 (a) Linear twist function $f(y)$ given in Section 6.2.8, with $y_0 = 1/4$, $y_1 = 3/4$ and $k = l = 1$. Note the loss of differentiability at $y = y_0, y_1$. (b) Everywhere-differentiable twist function $f(y)$ given in Section 6.2.6 with parameters as in (a). The function is differentiable in $[y_0, y_1]$.

and in this case we have

$$\alpha = \frac{k}{y_1 - y_0} = \frac{df}{dy} \quad \forall y \in (y_0, y_1),$$

$$\beta = \frac{l}{x_1 - x_0} = \frac{dg}{dx} \quad \forall x \in (x_0, x_1).$$

We show this linear twist function in figure 6.5(a). Note that while the derivative of f (resp. g) is identical on the interior of $[y_0, y_1]$ (resp. $[x_0, x_1]$), f (resp. g) is not differentiable at the endpoints $y = y_0, y_1$ (resp. $x = x_0, x_1$) since the identity maps outside the annuli have the derivative zero.

6.2.9 More general twists

It is possible for the functions f and g to have even more generality and for the following theorems to still hold. For example it is possible for f and g to have finitely many discontinuities or points of inflexion. We note here in particular the work of Nicol (1996b), in which certain toral linked twist maps with countably many discontinuities are shown to have the Bernoulli property, to have infinite entropy and to be stable with respect to a class of random perturbations; and Nicol (1996a), in which a Bernoulli toral linked twist map is constructed which has positive Lyapunov exponents only on a set of zero measure. These scenarios are unlikely to occur in applications however. The crucial feature here the twist functions must possess is monotonicity. Without this the following theorems, as they stand, are not valid.

In fluid mechanical applications it may be rare to see monotonic twist functions. Often physical constraints such as non-slip boundary conditions result in a non-monotonic velocity profile. In Chapter 9 we discuss some of the issues connected with this.

6.3 The ergodic partition for smooth toral linked twist maps

6.3.1 Co-rotating smooth toral linked twist maps

Throughout Section 6.3 we assume we have a smooth map on the whole torus – that is, twists given by the general smooth twists defined in Section 6.2.6. The following result was proved in Burton & Easton (1980).

Theorem 6.3.1 (Burton & Easton (1980)) *For a toral linked twist map H composed of (k, α) and (l, β) smooth twists in which k and l have the same sign, then $R = P \cup Q$ has an ergodic partition (that is, R is a union of (possibly countably many) ergodic components of positive measure).*

To prove this result we follow the approach of Burton & Easton (1980), using the everywhere-differentiable form of the twist in Section 6.2.6. This means we can use Pesin's results directly without needing to appeal to the Katok–Strelcyn version for maps with singularities. The case with singularities will be discussed in the following section. We proceed using the following lemmas. The basic idea of the proof is to define a set for which we can show that the Lyapunov exponents are non-zero (for all tangent vectors). Then Pesin theory (see Theorem 5.3.7) can be applied to deduce an ergodic partition. To complete the proof we must show that this set contains almost every point in R (that is, all points up to a set of measure zero). More formally, let

$$\Lambda = \{z = (x, y) \in R : \chi^{+}(z, v) \neq 0 \text{ for each } v \neq 0, v \in T_z R\}.$$

This is the set of all initial conditions whose trajectory gives rise to a non-zero Lyapunov exponent for any initial tangent vector. Pesin's results state that Λ has an ergodic partition. By considering the following sets and functions we construct a set of points with non-zero Lyapunov exponents – that is, a set of points contained in Λ. Note that Lyapunov exponents exist since the Oseledec condition **(OS)** (see Equation (5.22)) is trivially satisfied as our map is smooth and the torus compact.

Fix a number $\delta \in (0, 1)$ and define

$$A_1(\delta) = \{(x, y) \in P : y_0 + \delta(y_1 - y_0)/2 \leq y \leq y_1 - \delta(y_1 - y_0)/2\},$$

$$A_2(\delta) = \{(x, y) \in Q : x_0 + \delta(x_1 - x_0)/2 \leq x \leq x_1 - \delta(x_1 - x_0)/2\},$$

$$A_3(\delta) = \{(x, y) \in A_1(\delta) : H(x, y) \in A_2(\delta)\}.$$

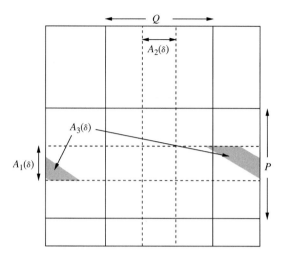

Figure 6.6 The strips $A_1(\delta)$ and $A_2(\delta)$, together with the set $A_3(\delta)$, shaded. Points in $A_3(\delta)$ map into $A_2(\delta)$ under a single iteration of H.

The sets $A_1(\delta)$ and $A_2(\delta)$ are strips of width $(y_1 - y_0)(1 - \delta)$ in P and $(x_1 - x_0)(1 - \delta)$ in Q respectively. Here we note that although for the smooth toral linked twist map we have $\alpha, \beta = 0$, the derivatives f' and g' are bounded away from 0 for points in $A_1(\delta)$ and $A_2(\delta)$ for a fixed $\delta > 0$. The set $A_3(\delta)$ contains all points in $A_1(\delta)$ which map into $A_2(\delta)$ under a single iterate of H (of course this is the same as the sets of points in $A_1(\delta)$ which map into $A_2(\delta)$ under a single iteration of F, since G leaves the x-coordinate invariant). See Figure 6.6.

Define the function $\chi_{A_3(\delta)}$ to be the characteristic function[3] for $A_3(\delta)$, so

$$\chi_{A_3(\delta)} : R \to \{0, 1\}$$

$$\chi_{A_3(\delta)}(x, y) = \begin{cases} 1 & \text{if } (x, y) \in A_3(\delta) \\ 0 & \text{if } (x, y) \notin A_3(\delta). \end{cases}$$

Finally define

$$A_4(\delta) = \{(x, y) \in R : \lim_{n \to \infty} \frac{1}{n} \sum_{i=0}^{n-1} \chi_{A_3(\delta)}(H^i(x, y)) \geq \delta\},$$

so $A_4(\delta)$ is the set of points in R whose trajectory under iterates of H lands in $A_3(\delta)$ with frequency at least δ (that is, returns on average at least once every $1/\delta$ iterations). Now we claim that points in $A_4(\delta)$, for any fixed δ, have non-zero Lyapunov exponents.

---

[3] We have previously encountered the characteristic function in Section 3.7.

Lemma 6.3.1 (Burton & Easton (1980)) *If $z = (x, y) \in A_4(\delta)$ and if $v \in T_z R$*
($v \neq 0$), then $\chi^+(z, v) \neq 0$. Consequently $A_4(\delta) \subset \Lambda$.

Proof For any point $z = (x_0, y_0) \in A_4(\delta)$, denote the iterates of z under H by
$H^i(z) = z_i = (x_i, y_i)$. Also we denote $F(z) = (x_0', y_0)$, and similarly $F(H^i(z)) =$
(x_i', y_i) (note that $x_i = x_i'$ if $H^i(z) \in R\backslash P$). Then

$$DH|_z = DG|_{F(z)} DF|_z = \begin{pmatrix} 1 & 0 \\ g'(x_0') & 1 \end{pmatrix} \begin{pmatrix} 1 & f'(y_0) \\ 0 & 1 \end{pmatrix} \qquad (6.1)$$

if $z \in P$ and $F(z) \in Q$. If $z \notin P$ then the matrix DF is the identity matrix, and
similarly if $F(z) \notin Q$ then DG is the identity matrix. Therefore, using the chain
rule to compute the Jacobian along a trajectory, we have

$$DH^n|_z = DH|_{H^{n-1}(z)} DH|_{H^{n-2}(z)} \cdots DH|_{H(z)} DH|_z$$

and so $DH^n|_z$ is a matrix of the form

$$DH^n|_z = \begin{pmatrix} 1 & a_{n-1} \\ b_{n-1} & 1 + a_{n-1}b_{n-1} \end{pmatrix} \cdots \begin{pmatrix} 1 & a_0 \\ b_0 & 1 + a_0 b_0 \end{pmatrix},$$

where $a_i = f'(y_i)$ if $z_i \in P$ and $a_i = 0$ if $z_i \in R\backslash P$. Similarly $b_i = g'(x_i')$ if
$F(z_i) \in Q$ and $b_i = 0$ if $F(z_i) \in R\backslash Q$. Hence each of the a_i, $b_i \geq 0$ since we
have chosen $k, l > 0$. In particular, for a fixed $\delta \in (0, 1)$, when $z_j \in A_3(\delta)$, a_j,
$b_j \geq \kappa > 0$ where

$$\kappa = \min \big\{ f'(y) : y \in [y_0 + \delta(y_1 - y_0)/2, y_1 - \delta(y_1 - y_0)/2],$$
$$g'(x) : x \in [x_0 + \delta(x_1 - x_0)/2, x_1 - \delta(x_1 - x_0)/2] \big\}.$$

Note that κ is strictly positive as we have fixed $\delta > 0$.

To compute Lyapunov exponents we must evaluate $\|DH^n|_z v\|$ for a general
vector $v = (v_1, v_2)$ in tangent space $T_z R$. For the rest of this proof, for ease
of notation, we will drop the subscript z and write $\|DH^n v\|$. The idea behind
the argument is effectively an invariant cone field argument, although here
we give the details of computing Lyapunov exponents explicitly. We begin by
demonstrating that the first quadrant of tangent space is invariant and expanding,
in the sense that tangent vectors in this quadrant remain in the quadrant, and
have increasing Euclidean norm. The same is true of the third quadrant, and
these form the expanding cone in tangent space.

We first suppose $v_1, v_2 \geq 0$. Then since the a_i, $b_i \geq 0$, both components
of $\|DH^k v\|$ are greater than or equal to zero for each k, and we may show as
follows that $\|DH^n v\|$ is a non-decreasing function of n. (Note that this statement
is not true for the counter-rotating case $kl < 0$.) First observe the growth of

$\|DH^n v\|$ whenever an iterate falls into $A_3(\delta)$. If $H^k(z) \in A_3(\delta)$ for some k then since $a_k, b_k \geq \kappa$ and $v_1, v_2 \geq 0$, we have for $DH^k v = (w_1, w_2)$,

$$
\begin{aligned}
\|DH^{k+1}v\| &= \left\| \begin{pmatrix} 1 & a_k \\ b_k & 1+a_k b_k \end{pmatrix} DH^k v \right\| \\
&= \left\| \begin{pmatrix} w_1 + a_k w_2 \\ b_k w_1 + (1 + a_k b_k) w_2 \end{pmatrix} \right\| \\
&= \sqrt{w_1^2(1+b_k^2) + w_2^2(a_k^2 + (1+a_k b_k)^2) + w_1 w_2 (2a_k + 2b_k(1+a_k b_k))} \\
&\geq \sqrt{w_1^2(1+\kappa^2) + w_2^2(1 + 3\kappa^2 + \kappa^4) + w_1 w_2(2\kappa + 2\kappa(1+\kappa^2))} \\
&\geq \sqrt{(w_1^2 + w_2^2)(1 + \kappa^2)} \\
&= \sqrt{1 + \kappa^2} \|DH^k v\|.
\end{aligned}
$$

If this growth occurred at every iteration the proof would be finished. We now show that this growth occurs *sufficiently often* to deduce the result. Recall that $A_4(\delta)$ contains points whose trajectory lands in $A_3(\delta)$ with frequency at least δ. That means that in n iterations, δn of them land in $A_3(\delta)$ *on average*. Although this does not mean that in *every* n iterations we have at least δn landing in $A_3(\delta)$, a simple combinatorial argument will show that if we choose n sufficiently large, at least $\delta n/2$ iterations land in $A_3(\delta)$. In other words, there exists an n sufficiently large that $H^k(z) \in A_3(\delta)$ for at least l integers k between 0 and $n-1$, where l is such that $l > \delta n/2$. Hence

$$
\|D_z H^n v\| \geq (\sqrt{(1+\kappa^2)})^l \|v\| > (1 + \kappa^2)^{n\delta/4} \|v\|,
$$

and therefore

$$
\frac{1}{n} \log \|D_z H^n v\| > \frac{\delta}{4} \log(1 + \kappa^2) + \frac{1}{n} \log \|v\|
$$

and so

$$
\chi^+(z, v) = \lim_{n \to \infty} \frac{1}{n} \log \|D_z H^n v\| > \frac{\delta}{4} \log(1 + \kappa^2) > 0.
$$

This has shown the result only for vectors v for which $v_1, v_2 \geq 0$. For complete nonuniform hyperbolicity we need it to be true for all vectors in $T_z R$. Lyapunov exponents have the linearity property $\chi^+(z, sv) = \chi^+(z, v)$ for any real $s \neq 0$, and so setting $s = -1$ gives $\chi^+(z, v) > 0$ for all vectors v for which $v_1, v_2 \leq 0$ (alternatively minor changes to the above argument will suffice). Thus to show $\chi^+(z, v) \neq 0$ for all $v \in T_z R$ we just have to demonstrate the result for $v_1 < 0$, $v_2 > 0$ (the cases $v_1, v_2 < 0$ and $v_1 > 0, v_2 < 0$ follow by the linearity property with $s = -1$). If for some n, $D_z H^n v$ has both components non-negative, then we can apply the argument above to show that $\chi^+(z, v) > 0$.

The only remaining possibility is that $D_z H^n v$ has a negative first component and a positive second component for all $n > 0$. Consider $w \in T_z R$ with $w_1 < 0$ and $w_2 > 0$. Now letting $z = (x_0, y_0)$ and $F^{-1}(z) = (x_0', y_0')$ we have

$$D_z H^{-1} w = \begin{pmatrix} 1 + f'(y_0) g'(x_0') & -f'(y_0) \\ -g'(x_0') & 1 \end{pmatrix} \begin{pmatrix} w_1 \\ w_2 \end{pmatrix}$$

also has first component negative and second component positive. Moreover, similarly to the above,

$$\|D_z H^{-1} w\| \geq \sqrt{1 + \kappa^2} \|w\|$$

and then a similar argument gives $\chi^+(z, v) < -(\delta/4) \log(1 + \kappa^2) < 0$. □

The above technique of computing Lyapunov exponents for linked twist maps illustrates a fundamental idea. Equation (6.1) gives an expression for the Jacobian of the map which is valid only for iterates which begin in the intersection S and fall into S after the first twist F. If this is the case for every iterate (as must happen, for example, if $x_0 = y_0 = 0$, $x_1 = y_1 = 1$, and we have a toral automorphism – see Example 5.2.4), then the work is simple. The logarithms of the eigenvalues of the matrix DH are also the Lyapunov exponents for each point on the torus, and we have uniform hyperbolicity throughout. The complication here stems from the fact that not every iterate falls into S every time. The above lemma shows, in the co-rotating case, that *providing* each trajectory falls into S *sufficiently often*, then the contribution from DH ensures non-zero Lyapunov exponents. However the fact that the number of iterates between visits to S may differ between individual points in R means that we have nonuniform hyperbolicity.

> **Key point:** For the smooth co-rotating toral linked twist map, Lyapunov exponents are non-zero for all points falling into the overlap region sufficiently often. It is possible to prove this because tangent vectors are expanded by both horizontal and vertical twist maps every time an iterate hits the intersection, and this expansion is not undone during times away from the overlap. Iterates visit the overlap sufficiently frequently to allow Lyapunov exponents to be bounded away from zero.

To show that the whole region R has an ergodic partition, we must show that Λ has full measure in R. Since $A_4(\delta) \subset \Lambda \subset R$ for each δ, in order to prove this, it is enough to prove that the set

$$\bigcup \{A_4(\delta) : 0 < \delta < 1\}$$

has full measure in R. Before proving this result, we give two lemmas which will be useful here and in later work. The first says that the orbit of almost every point in R must hit S under forward iteration, and the second says that almost all orbits that intersect S (or in fact any measurable set) under forward iteration continue to do so, with positive frequency (a result similar to the Poincaré Recurrence Theorem, Theorem 3.5.1).

Lemma 6.3.2 (Burton & Easton (1980))

   (i) *For almost every* $z \in P$, $F^n(z) \in S$ *for some* $n > 0$.
  (ii) *For almost every* $z \in Q$, $G^n(z) \in S$ *for some* $n > 0$.
 (iii) *For almost every* $z \in R$, $H^n(z) \in S$ *for some* $n > 0$.

Proof (i) Suppose $z \in P$. Then $F^n(z) \in S$ for some $n > 0$ unless $F^n(z) \in P \backslash S$ for all $n > 0$. However, since F is a rigid rotation on each circle $y = $ constant, the only points for which this occurs form a set of zero measure, namely, a subset of the set $\{(x, y) | f(y) \in \mathbb{Q}\}$ (recall that a one-to-one function takes a countable set of points onto a countable set of points). The proof of (ii) is precisely analogous, and (iii) follows as a direct consequence of (i) and (ii).   □

Lemma 6.3.3 (Burton & Easton (1980)) *Let V be a measurable subset of R and define*[4]

$$J(V) = \{z \in R | H^n(z) \in V \text{ for some } n \geq 0\},$$

that is, the set of points whose orbit hits V under forward iteration, and define

$$Z(V) = \{z \in R | \lim_{n \to \infty} \frac{1}{n} \sum_{i=0}^{n-1} \chi_V(H^i(z)) = 0\},$$

that is, the set points who hit V with zero frequency on average. Then

$$\mu(J(V) \cap Z(V)) = 0.$$

In other words, $J(V)$ and $Z(V)$ coincide only on a set of measure zero, so that almost all orbits that hit V under forward iteration continue to hit V with positive frequency in n.

---

[4] In fact, as in Burton & Easton (1980), this lemma holds for any compact metric space with a Borel measure-preserving homeomorphism. Note the relation to Kac's theorem, which gives the average return time to a measurable set under an ergodic transformation (Pollicott & Yuri (1998)).

Proof First consider the set $Z(V) \cap V$. We have

$$\mu(Z(V) \cap V) = \int_{Z(V)} \chi_V(z) d\mu$$

$$= \int_{Z(V)} \lim_{n \to \infty} \frac{1}{n} \sum_{i=0}^{n-1} \chi_V(H^i(z)) d\mu$$

$$= 0.$$

Here the first equality follows from the definition of χ_V, the second is an application of the Birkhoff ergodic theorem (Theorem 3.5.2), and the third follows from the definition of $Z(V)$. Now define

$$Z_k(V) = \{z \in Z(V) | H^k(z) \in V \text{ for some } k \geq 0 \text{ and } H^i(z) \notin V \text{ for } 0 \leq i < k\},$$

that is, the set of points in $Z(V)$ which take exactly k iterations to first hit V. Now since the kth iterate of the set $Z_k(V)$ contains points which by definition lie in V we have

$$H^k(Z_k(V)) \subset Z(V) \cap V$$

and so

$$\mu(Z_k(V)) = \mu(H^k(Z_k(V)))$$

$$\leq \mu(Z(V) \cap V)$$

$$= 0,$$

where the first equality follows since H is a μ-preserving map. However we also have

$$J(V) \cap Z(V) = \bigcup_{k \geq 0} Z_k(V)$$

since the union of the $Z_k(V)$ contains points in $Z(V)$ which hit V under some forward iterate of H. Finally, a countable union of sets of zero measure is also a set of zero measure, and so

$$\mu(J(V) \cap Z(V)) = 0.$$

$$\square$$

Now we are in a position to prove the following:

Lemma 6.3.4 (Burton & Easton (1980)) $\bigcup\{A_4(\delta) : 0 < \delta < 1\}$ *has full measure in R.*

Proof Choose a monotonic decreasing sequence $\{\delta_k\}$ converging to zero and define

$$U_k = J(A_3(\delta_k))\backslash Z(A_3(\delta_k)).$$

Here we take J and Z as in the previous lemma, and choose $V = A_3(\delta_k)$. The orbit of a point z' in U_k hits $A_3(\delta_k)$ with some positive frequency $\phi > 0$ (U_k is effectively all the points that hit $A_3(\delta_k)$ at least once, minus all the points that hit it with frequency zero). Hence the orbit of z' hits $A_3(\delta_{k'})$ with frequency at least $\delta_{k'}$ for each $\delta_{k'} < \phi$. Therefore z' belongs to $A_4(\delta_{k'})$ and so $U_k \subset \bigcup_k \{A_4(\delta_k) : 0 < \delta_k < 1\}$.

Now set $U = \bigcup_k \{U_k : k \geq 0\}$. Since $U_k \subset \bigcup_k \{A_4(\delta_k) : 0 < \delta_k < 1\}$, we have $U \subset \bigcup_k \{A_4(\delta_k) : 0 < \delta_k < 1\}$ also. This gives us

$$\mu(\cup_k\{A_4(\delta_k)\}) \geq \mu(U)$$

$$= \mu(\cup_k\{U_k\})$$

$$= \mu(\cup_k\{J(A_3(\delta_k))\backslash Z(A_3(\delta_k))\})$$

$$= \mu(\cup_k\{J(A_3(\delta_k))\}) - \mu(\cup_k\{J(A_3(\delta_k)) \cap Z(A_3(\delta_k))\})$$

$$= \mu(\cup_k\{J(A_3(\delta_k))\}) - 0 \text{ (by Lemma 6.3.3)}$$

$$= \mu(\cup_k\{J(A_3(\delta_k))\})$$

where all the unions are countable unions taken over $0 < \delta_k < 1$. Therefore it is sufficient to show $\bigcup_{k \geq 0}\{J(A_3(\delta_k))\}$ has full measure in R in order to prove that $\bigcup_{k \geq 0}\{A_4(\delta_k)\}$ has full measure. But $\bigcup_{k \geq 0}\{J(A_3(\delta_k))\}$ is by definition the set of points which iterate into S for some iterate, and by Lemma 6.3.2 this is a set of full measure.   □

Proof of Theorem 6.3.1 Lemma 6.3.1 shows that there is a set of points in the domain of the toral linked twist map with non-zero Lyapunov exponents, and Lemma 6.3.4 shows that this set contains almost every point in the domain. Then we can appeal to Theorem 5.3.7 to complete the proof.   □

Key point: Almost every point falls into the overlap, and of the points which fall into the overlap, almost all of them do so with positive frequency. These facts allow us to deduce that for the smooth co-rotating toral linked twist map, almost all points in the union of the annuli give rise to a non-zero Lyapunov exponent.

Example 6.3.1 *Figure 6.7 shows a co-rotating system, for a smooth version of the twist function, with $x_0 = y_0 = 1/4$, $x_1 = y_1 = 3/4$, and $k = l = 1$,*

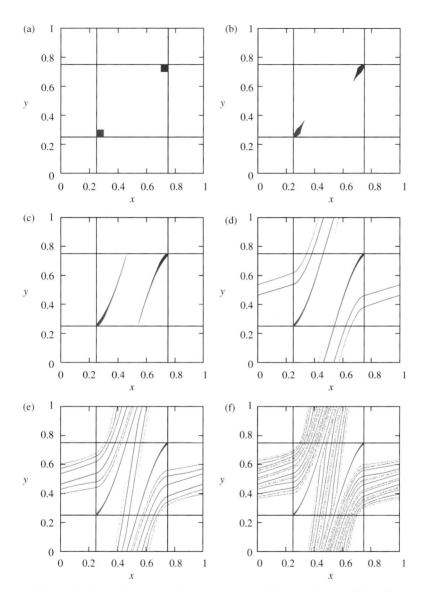

Figure 6.7 Successive iterates of the co-rotating toral linked twist map (6.1), with $x_0 = y_0 = 1/4$, $x_1 = y_1 = 3/4$, and $k = l = 1$ for the everywhere-differentiable twist map. Initially (in the top left picture) we have blobs at $[0.25, 0.3] \times [0.25, 0.3]$ and $[0.7, 0.75] \times [0.7, 0.75]$. The next five pictures are the images of these blobs under the first five iterates of the map. It is interesting to compare the behaviour of this map with that of the Arnold Cat Map (see figure 3.4).

for successive iterates of two initial blobs, which are the squares $[0.25, 0.3] \times$
$[0.25, 0.3]$ *and* $[0.7, 0.75] \times [0.7, 0.75]$. *This shows the stretching and subsequent*
intermingling of trajectories within the intersection S.

6.3.2 Counter-rotating smooth toral linked twist maps

The above arguments only hold for the co-rotating case. In the counter-rotating
version, we cannot apply the same argument to show that that all vectors in
tangent space are expanded (or contracted) by the Jacobian matrix at each
iteration of the map, since, for example, for a vector $w = (v_1, v_2)$ with $v_1 > 0$,
$v_2 > 0$, $\|D_z H^n w\|$ may not be a non-decreasing function of n. Later, in section
6.4.2, we will see how to construct non-zero Lyapunov exponents in this case,
but this requires stronger conditions on the twist functions.

Example 6.3.2 *Figure 6.8 shows a counter-rotating system, for the smooth*
version of the twist function, with $x_0 = y_0 = 1/4$, $x_1 = y_1 = 3/4$, *and*
$k = l = 1$, *for iterates of two initial blobs, which are the squares* $[0.25, 0.3] \times$
$[0.25, 0.3]$ *and* $[0.5, 0.55] \times [0.5, 0.55]$. *Whilst the central cross-shaped region*
is a region on which we appear to have ergodicity, near the boundaries of R we
do not.

Key point: Smooth counter-rotating linked twist maps are not open to the same
analysis as the smooth co-rotating version. This is because expansion due to a
horizontal twist may be undone by contraction due to a vertical twist, and so expansion under the composed linked twist map cannot be guaranteed. Such expansion
requires extra conditions on the strengths of the twists.

6.4 The ergodic partition for toral linked twist maps with singularities

Throughout Section 6.4 we assume we have the non-smooth twists defined in
Section 6.2.7. Referring back to the Katok–Strelcyn conditions in Section 5.3
we see that toral linked twist maps with singularities fit the required framework.
Our manifold (recall Figure 6.1) M is made up of rectangles S, $P \backslash S$ (possibly
two disconnected components) and $Q \backslash S$ (also possibly two components) 'glued'
together along the lines contained in the set $G = \partial S$. Then the (open dense) set
$V = M \backslash G = M \backslash \partial S$. The open subset $N \subset V$ can be given to be $N = \text{int } V =$
$\text{int } (M \backslash G)$. Now H is a C^2 measure-preserving mapping on N, since we have

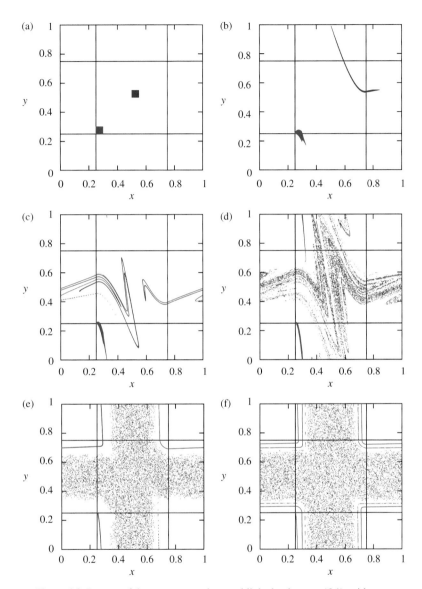

Figure 6.8 Iterates of the counter-rotating toral linked twist map (6.1), with $x_0 = y_0 = 1/4$, $x_1 = y_1 = 3/4$, and $k = l = 1$ for the smooth twist map. Initially (in the top left picture) we have blobs at $[0.25, 0.3] \times [0.25, 0.3]$ and $[0.7, 0.75] \times [0.7, 0.75]$. The next five pictures are the images of these blobs under the iterates 2, 5, 10, 30, 100 of the map.

removed the points on which H is not differentiable. Finally, sing$H = M \backslash N = \partial R \cup \partial S$.

6.4.1 Co-rotating toral linked twist maps with singularities

In this section we give the corresponding result to Theorem 6.3.1 for toral linked twist maps with singularities. The computation of Lyapunov exponents is simplified here as we have derivatives of the twists bounded away from zero, having removed the singularity set sing$H$.

Theorem 6.4.1 (Wojtkowski (1980), Przytycki (1983)) *Let H be a toral linked twist map composed of (k, α) and (l, β) twists in which k and l have the same sign, and $\alpha\beta > 0$. Then H is a union of (possibly countably many) ergodic components of positive measure.*

To apply the Katok–Strelcyn theorem we must show that H satisfies the Katok–Strelcyn conditions (see Section 5.3). In addition we must show that Lyapunov exponents for H exist almost everywhere (i.e., show that the Oseledec condition holds), and that these exponents are non-zero. In fact the Katok–Strelcyn conditions for H hold very easily:

Lemma 6.4.1 *H satisfies* **(KS1)**.

Proof $B(\mathrm{sing}H, \epsilon) = B(\partial R \cup \partial S, \epsilon)$. An ϵ-neighbourhood around $\partial R \cup \partial S$ consists of 4 rectangles around the top and bottom edges of P, and the left and right edges of Q, each of width 2ϵ and length 1. Thus $\mu(B(\mathrm{sing}H, \epsilon)) = 4 \times 2\epsilon$, and so Equation (5.24) is easily satisfied with $C_1 = 8$ and $a = 1$. $\qquad\square$

This also implies that $\mu(\mathrm{sing}H) = 0$.

Assumption 6.4.1 *H satisfies* **(KS2)**.

This is not a severe restriction, in particular for applications. For example, if the twist functions f and g are polynomials. Equation (5.25) is trivially satisfied. Any continuous function on a compact interval can be approximated arbitrarily closely by some polynomial function (the Weierstrass approximation theorem). A function for which the second derivative grows faster near a singularity, for example, $f(y) = \sqrt{y}$ near $y = 0$, also straightforwardly satisfies the condition. The condition fails for function whose second derivative grows faster than any polynomial, for example, $f(y) = e^{1/y}$ near $y = 0$.

To compute Lyapunov exponents of H we consider the *first return map* to S; that is, the map H_S defined earlier. It is simple to check the Oseledec conditions for both maps H_S and H:

Lemma 6.4.2 *H and H_S satisfy the Oseledec condition* **(OS)**.

Proof

$$\int_S \log^+ \|DH_S(x)\| d\mu \le \int_R \log^+ \|DH(x)\| d\mu$$

$$= \int_V \log^+ \|DH(x)\| d\mu$$

$$< \infty$$

where the first inequality follows because $S \subseteq R$, the equality follows since $\mu(R \backslash V) = \mu(\text{sing} H) = 0$, and the second inequality because H is C^2 on $V = \text{int} R \backslash \partial S$. $\qquad \square$

Next we compute bounds on the Lyapunov exponents for H_S.

Lemma 6.4.3 *Lyapunov exponents for H_S are non-zero.*

Proof This follows essentially the same argument as Lemma 6.3.1. We have for each $z \in S$

$$DH_S = DG_S DF_S = \begin{pmatrix} 1 & 0 \\ \tilde{\beta} & 1 \end{pmatrix} \begin{pmatrix} 1 & \tilde{\alpha} \\ 0 & 1 \end{pmatrix} = \begin{pmatrix} 1 & \tilde{\alpha} \\ \tilde{\beta} & 1 + \tilde{\alpha}\tilde{\beta} \end{pmatrix}$$

where $\tilde{\alpha} \ge m\alpha > 0$, and $\tilde{\beta} \ge m'\beta > 0$, with m and m' as in the definitions of F_S and G_S (Definitions 6.2.3 and 6.2.4). Recall that m and m' depend on the point z. Let $\kappa = \min_{z \in S}\{m\alpha, m'\beta\} > 0$.

Then by the same computation as earlier for $w = (v_1, v_2)$ where $v_1, v_2 > 0$,

$$\|DH_S w\| = \left\| \begin{pmatrix} 1 & \tilde{\alpha} \\ \tilde{\beta} & 1 + \tilde{\alpha}\tilde{\beta} \end{pmatrix} \begin{pmatrix} v_1 \\ v_2 \end{pmatrix} \right\|$$

$$\ge \sqrt{1 + \kappa^2} \|w\|.$$

Because H_S is the first return map to S this is valid for each iterate, and so

$$\|DH_S^n w\| \ge (1 + \kappa^2)^{n/2} \|w\|$$

and hence

$$\chi^+(z, w) = \lim_{n \to \infty} \frac{1}{n} \log \|D_z H_S^n(w)\| \ge \frac{1}{2} \log(1 + \kappa^2) > 0.$$

For other vectors in tangent space we proceed as in Lemma 6.3.1. $\qquad \square$

Working with H_S instead of H simplifies the computation of Lyapunov exponents as we fall into the intersection S under *every* iterate of H_S. This means that in fact H_S is *uniformly* hyperbolic. Hence any points for which $H_S \equiv H$ (that is, any points falling into the intersection on every iterate) form a uniformly hyperbolic set. This set is a set of measure zero however, and forms the horseshoe constructed in Chapter 4. However, we would like to show the (non-uniform) hyperbolicity of H itself, and so finally we conclude that Lyapunov exponents for H are also non-zero.

Lemma 6.4.4 *Lyapunov exponents for H are non-zero almost everywhere.*

Proof By Lemma 6.3.2 almost every $z \in R$ hits S. Moreover almost every $z \in S$ returns to S with positive frequency $\delta_z > 0$ by Lemma 6.3.3. Recalling that $H^j(z) = H_S^n(z)$ where $l = \sum_{i=0}^{n-1}(m_i + m'_i - 1)$ we have, using a similar argument to Lemma 6.3.1, that for j sufficiently large, at least δ_z iterates of H^j hit S. Therefore $n > j\delta_z/2$. Then using Lemma 6.4.3 we have

$$
\frac{1}{j} \log \|DH^j v\| = \frac{1}{j} \log \|DH_S^n v\|
$$

$$
> \frac{1}{j} \log \|DH_S^{j\delta_z/2} v\|
$$

$$
\geq \frac{1}{j} \log(1 + \kappa^2)^{j\delta_z/4} \|v\|
$$

$$
= \frac{\delta_z}{4} \log(1 + \kappa^2) + \frac{1}{j} \log \|v\|.
$$

Therefore

$$
\lim_{j \to \infty} \frac{1}{j} \log \|DH^j(v)\| > \frac{\delta_z}{4} \log(1 + \kappa^2) > 0.
$$

$\square$

Proof of Theorem 6.4.1 Lemma 6.4.2 shows that Lyapunov exponents for H exist, and Lemma 6.4.4 shows that they are non-zero. Lemma 6.4.1 and Assumption 6.4.1 show that H satisfies the Katok–Strelcyn conditions. Then Theorem 5.4.1 completes the proof. $\square$

Example 6.4.1 *Figure 6.9 shows such a co-rotating system, now for the linear twist function of Section 6.2.8, with $x_0 = y_0 = 1/4$, $x_1 = y_1 = 3/4$, and now for $k = l = 2$, for successive iterates of two initial blobs, which are the squares $[0.25, 0.3] \times [0.25, 0.3]$ and $[0.7, 0.75] \times [0.7, 0.75]$. This shows the stretching of trajectories within the intersection S, and also the complicated (non-differentiable) angular structure of orbits, as line segments lie over the boundary of S. It is this geometric structure which we will have to analyse in the next chapter to strengthen the ergodic property result into mixing results. Note that already by the fifth iteration (in the final picture), a large majority of the domain R appears to be mixed. A few iterates later the region of mixing has spread to the entirety of R.*

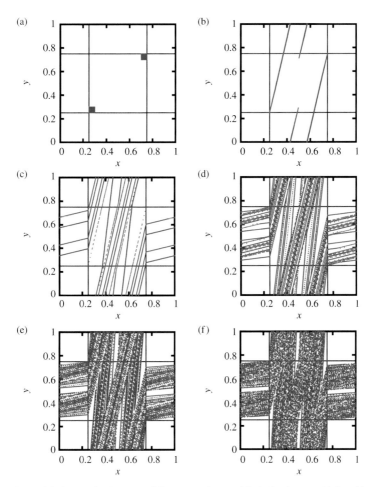

Figure 6.9 Successive iterates of the co-rotating toral linked twist map (6.1), with $x_0 = y_0 = 1/4, x_1 = y_1 = 3/4$, and $k = l = 2$. Initially (in the top left picture) we have blobs at $[0.25, 0.3] \times [0.25, 0.3]$ and $[0.7, 0.75] \times [0.7, 0.75]$. The next five pictures are the images of these blobs under the first five iterates of the map.

6.4.2 Counter-rotating toral linked twist maps with singularities

The result for counter-rotating toral LTMs has been proved by Wojtkowski (1980), who uses techniques based on the approach of Anosov and Sinai for showing ergodicity of Anosov diffeomorphisms. These techniques predate the work of Pesin and involve constructing expanding and contracting subspaces directly. A major technical obstacle arises in this approach, namely, showing the absolute continuity of stable and unstable foliations. This is necessary to overcome the problem that stable and unstable manifolds at a point z, being defined in terms of infinite forward and backward iterations respectively, may not depend differentiably on z. An excellent discussion of this issue can be found in Brin & Stuck (2002). Using Pesin theory and the Katok–Strelcyn theory allows us to avoid these difficulties. The following proof is based on Przytycki (1983).

Theorem 6.4.2 (Wojtkowski (1980), Przytycki (1983)) *For a toral linked twist map H composed of (k, α) and (l, β) twists in which k and l have opposite signs, and*

$$|\alpha\beta| > 4, \tag{6.2}$$

then H is a union of (possibly countably many) ergodic components of positive measure.

Note that smooth counter-rotating toral linked twist maps cannot satisfy the condition (6.2) and so as discussed earlier need not possess an ergodic partition. A simple example from Wojtkowski (1980) demonstrates that some condition on α and β is necessary.

Example 6.4.2 *Choose the counter-rotating toral linked twist map given by $x_0 = y_0 = 0$, $x_1 = y_1 = 1$ (so that we have a toral automorphism), with linear twists. Take $k = 1$ and $l = -1$ so that $\alpha = 1$ and $\beta = -1$. Then $F(x,y) = (x+y, y)$, $G(x,y) = (x, y-x)$ give $H(x,y) = (x+y, -x)$. Since in this case the intersection S is the whole torus $\mathbb{T}^2$, the orbit of any initial condition (x, y) is*

$$(x, y) \to (x+y, -x) \to (y, -x-y) \to (-x, -y)$$
$$\to (-x-y, x) \to (-y, x+y) \to (x, y)$$

and so every point is a period-6 point. Thus we have no ergodic behaviour in this system.

The proof of Theorem 6.4.2 follows the same structure as the co-rotating case (in particular the arguments for the Katok–Strelcyn conditions are identical), but we need new lemmas for computing Lyapunov exponents of H_S. We will use the method of invariant cones described in Section 5.5.

For this proof we assume for simplicity of notation that we have linear twists (see Section 6.2.8). This is justified since α and β are defined as the infimum and supremum respectively of derivatives, and so taking $f'(y) = \alpha$ for all y and $g'(x) = \beta$ for all x simply gives the case in which the minimum twist occurs at every point. We assume further that $\alpha = -\beta$. If not then a simple coordinate change $(x, y) \to (x, \sqrt{|\alpha/\beta|}y)$ will suffice. We take $\alpha > 0$ and $\beta < 0$, so that condition (6.2) becomes simply $\alpha > 2$. Finally, we assume that at least one of the inclusions $P \subseteq \mathbb{T}^2$ and $Q \subseteq \mathbb{T}^2$ is strict. If this is not the case, so that $P = Q = R = S = \mathbb{T}^2$ then we have a toral automorphism which is uniformly hyperbolic whenever the eigenvalues of $DH = DGDF$ are off the unit circle.

Similarly to the above (for points in P which map into Q) we have Jacobians

$$DF = \begin{pmatrix} 1 & \alpha \\ 0 & 1 \end{pmatrix} \quad \text{and} \quad DG = \begin{pmatrix} 1 & 0 \\ -\alpha & 1 \end{pmatrix}$$

and so

$$DGDF = \begin{pmatrix} 1 & \alpha \\ -\alpha & 1 - \alpha^2 \end{pmatrix}$$

where $\alpha > 2$. This matrix is hyperbolic and has eigenvalues

$$\lambda_{\pm} = \frac{2 - \alpha^2 \pm \sqrt{\alpha^4 - 4\alpha^2}}{2}.$$

Note that λ_- is the expanding eigenvalue ($|\lambda_-| > 1$), while $|\lambda_+| < 1$ is the contracting eigenvalue. The expanding eigenvector (u, v) satisfies

$$\frac{u}{v} = \frac{-\alpha}{2} + \sqrt{\left(\frac{\alpha}{2}\right)^2 - 1} = L.$$

At each point $z \in R$ we define the pair of cones $C(z)$ and $\tilde{C}(z)$ in tangent space $T_z R \equiv \mathbb{R}^2$ by

$$C(z) = \{(u, v) : L \leq u/v \leq 0\}$$

$$\tilde{C}(z) = \{(u, v) : u/v \geq L + \alpha\}.$$

Since each cone has the same definition for each $z \in R$, we simply denote them as C and $\tilde{C}$. Observing that since $\alpha > 2$, we have $-1 < L \leq 0$ and $1 < L + \alpha$, these are shown in Figure 6.10. We first show that these cones are 'pairwise' invariant (and eventually strictly invariant), and then that we have contraction or expansion for vectors in these cones. Since the cone C (resp. $\tilde{C}$) is the same at every point $z \in R$, we denote the cone field by C (resp. $\tilde{C}$) as well.

Lemma 6.4.5 *We have the inclusions*

$$DF(C) \subseteq \tilde{C},$$

and for an integer $m \geq 2$,

$$DF^m(C) \subset \tilde{C}.$$

Proof Suppose $w = (u, v) \in C$. Then $w' = (u', v') = DF^m w$, where

$$w' = \begin{pmatrix} 1 & m\alpha \\ 0 & 1 \end{pmatrix} \begin{pmatrix} u \\ v \end{pmatrix} = \begin{pmatrix} u + m\alpha v \\ v \end{pmatrix}$$

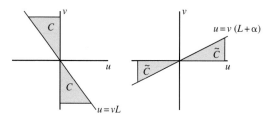

Figure 6.10 The cones $C = \{(u, v) : L \le u/v \le 0\}$ and $\tilde{C} = \{(u, v) : u/v \ge L+\alpha\}$ in tangent space.

with

$$\frac{u'}{v'} = \frac{u}{v} + m\alpha \ge L + m\alpha$$

so for $m = 1, (u', v') \in \tilde{C}$, and for $m \ge 2, (u', v') \in int(\tilde{C})$, giving the result. $\quad\square$

Lemma 6.4.6 *We have the inclusions*

$$DG(\tilde{C}) \subseteq C$$

and for an integer $m' \ge 2$,

$$DG^{m'}(\tilde{C}) \subset C.$$

Proof Suppose $w = (u, v) \in \tilde{C}$. Then $w' = (u', v') = DG^{m'}w$, where

$$w' = \begin{pmatrix} 1 & 0 \\ -m'\alpha & 1 \end{pmatrix} \begin{pmatrix} u \\ v \end{pmatrix} = \begin{pmatrix} u \\ v - m'\alpha u \end{pmatrix}$$

with

$$\frac{u'}{v'} = \frac{u/v}{1 - m'\alpha u/v} \le \frac{-1}{m'\alpha} \le 0$$

and

$$\frac{u'}{v'} = \frac{u/v}{1 - m'\alpha u/v}$$

$$\ge \frac{\alpha/2 + \sqrt{\alpha^2/4 - 1}}{1 - m'\alpha(\alpha/2 + \sqrt{\alpha^2/4 - 1})}$$

$$\ge \frac{\alpha/2 + \sqrt{\alpha^2/4 - 1}}{1 - \alpha^2/2 - \alpha\sqrt{\alpha^2/4 - 1}}.$$

For $m' = 1$, a little algebra will show that this last expression is equal to L, and so $(u', v') \in C$, while for $m' \geq 2$, we have $(u', v') \in \text{int}(C)$. $\qquad \square$

Corollary 6.4.1 *The cone field C is invariant and strictly invariant (unless $R = \mathbb{T}^2$) under H_S.*

Proof This follows from the previous two lemmas. For each $z \in R$, $H_S(z) = G^{m'} \circ F^m(z)$ (where m and m' depend on z as in Definitions 6.2.3 and 6.2.4). Thus C is mapped into $\tilde{C}$ under DF^m, which in turn is mapped back into C under $DG^{m'}$. The inclusion is strict as soon as one of m or m' is greater than one, which must occur since $S \neq T^2$. $\qquad \square$

Lemma 6.4.7 *For any $w \in C$,*

$$\|DG^{m'} DF^m w\| \geq \lambda \|w\|$$

where $\lambda > 1$, for any pair of positive integers m and m'.

Before proving this theorem, we note that we require to find the norm of a vector w. Earlier in this chapter we used the Euclidean norm. However there are many possible choices of norm. A vector norm simply has to satisfy three conditions of positivity, linearity and a triangle inequality (recall Definition 3.2.1). Here, for simplicity, we use the norm defined by taking the maximum of the moduli of each coordinate (the L_∞-norm):

$$\|(u, v)\| = \max\{|u|, |v|\}$$

It is simple to check that this satisfies the conditions to be a metric.

Proof of Lemma 6.4.7: Let $w = (u, v) \in C$. Then $\|w\| = |v|$ by the definition of the cone C. Recalling that if $w' = DG^{m'} DF^m w$, we have

$$w' = \begin{pmatrix} 1 & m\alpha \\ -m'\alpha & 1 - mm'\alpha^2 \end{pmatrix} \begin{pmatrix} u \\ v \end{pmatrix} = \begin{pmatrix} u + m\alpha v \\ -m'\alpha u + (1 - mm'\alpha^2)v \end{pmatrix}.$$

Then

$$\|w'\| \geq \| -m'\alpha u + (1 - mm'\alpha^2)v \|$$
$$= | -m'\alpha \frac{u}{v} + (1 - mm'\alpha^2) ||v|$$
$$\geq | -\alpha L + 1 - \alpha^2 ||v|$$
$$= |\lambda_-||v|$$
$$= \lambda \|w\|$$

where $\lambda = -\lambda_- > 1$. $\qquad \square$

Corollary 6.4.2 *The cone field C is expanding under H_S.*

Proof Again, since $H_S(z) = G^{m'} \circ F^m(z)$, this follows directly from the previous lemma. □

A contracting cone field can be found in an exactly analogous manner. Now we are in a position to apply Theorem 5.5.1.

Lemma 6.4.8 *Lyapunov exponents for H_S are non-zero μ-almost everywhere.*

Proof This follows directly from Corollaries 6.4.1, 6.4.2 and Theorem 5.5.1. □

Finally we deduce the corresponding result for H itself.

Lemma 6.4.9 *Lyapunov exponents for H are non-zero μ-almost everywhere.*

Proof The proof is identical to that of Lemma 6.4.4. □

Proof of Theorem 6.4.2 Lemma 6.4.2 shows that Lyapunov exponents for H exist, and Lemma 6.4.9 shows that they are non-zero. Lemma 6.4.1 and Assumption 6.4.1 show that H satisfies the KS conditions. Then Theorem 5.4.1 completes the proof. □

Key point: For a non-smooth linked twist map, it can be shown that singularities are sufficiently rare and well-behaved that Katok–Strelcyn theory may be applied. Then co-rotating toral linked twist maps, and counter-rotating toral linked twist maps with sufficiently strong twists, are nonuniformly hyperbolic, and so their domain can be partitioned into at most a countable number of components on which the dynamics is ergodic.

Example 6.4.3 *Figure 6.11 shows an identical system to Figure 6.9, but for the counter-rotating case $k = 2$, $l = -2$. Here again we see the whole domain gradually being filled. Note the acute angles in the images of the blobs. These will have a large part to play in the following chapter.*

Example 6.4.4 *Figure 6.12 shows a counter-rotating toral linked twist map which violates the conditions given by Equation (6.2) of Theorem 6.4.2. Here we do not see the blobs filling the whole region, but there is a large area of the domain left unmixed. Even after 300 iterates of the map there is still a large region untouched by iterates of the initial blobs. In fact the likely existence of KAM tori shows that this region will never be filled.*

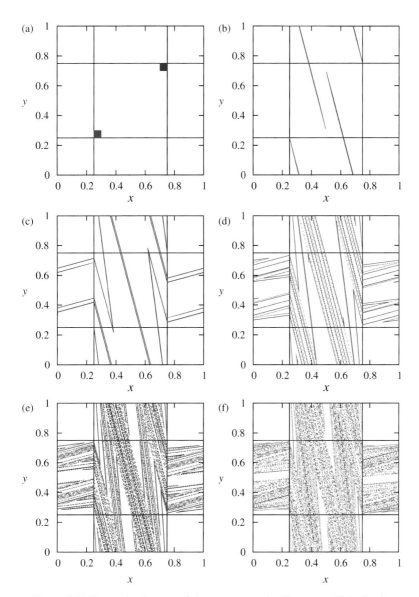

Figure 6.11 Successive iterates of the counter-rotating linear toral linked twist map, with $x_0 = y_0 = 1/4$, $x_1 = y_1 = 3/4$, and $k = 2, l = -2$. Initially (in the top left picture) we have blobs at $[0.25, 0.3] \times [0.25, 0.3]$ and $[0.7, 0.75] \times [0.7, 0.75]$. The next five pictures are the images of these blobs under the first five iterates of the map.

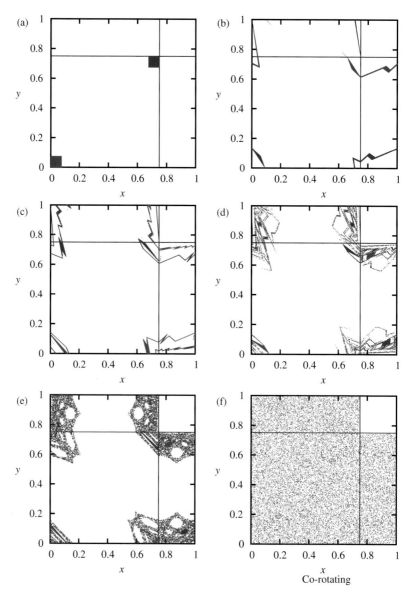

Figure 6.12 A counter-rotating toral linked twist map which violates the conditions of theorem 6.4.2. Here $x_0 = y_0 = 0$, $x_1 = y_1 = 3/4$, and $k = -l = 1$. The first five figures are 0, 5, 10, 20, 300 iterates of the initial blobs. The final picture, for comparison is an identical co-rotating map after 20 iterates. The difference between the behaviour of co- and counter-rotating tltm is striking.

6.5 Summary

The above theorems collect together the key techniques required to prove the existence of an ergodic partition for toral linked twist maps. The fundamental result underpinning it all is Theorem 5.3.7 from the field known as Pesin theory. To apply this theorem one needs to construct a set of points with non-zero Lyapunov exponents. In many systems this is straightforward (for example the toral automorphisms), but in the above we encountered some difficulties. First of all, for the smooth toral linked twist maps, we had a lower bound on the slope of the twist which was zero. This was overcome by taking a sequence of sets which accumulated on the annuli P and Q. A similar technique could be used if the functions f and g had inflexion points, but not turning points as we require monotonicity.

Another issue to arise is that of singularities in the map. These can be overcome by applying the extension of Pesin theory due to Katok *et al.* (1986). In the systems above we had a simple situation of singularities only at the boundaries ∂P and ∂Q. It would be straightforward however to extend the above to a situation with more singularities provided the Katok–Strelcyn and the Oseledec conditions hold.

Finally we note the technique of constructing the first return map H_S. This map is uniformly hyperbolic, and has an invariant set of measure zero which forms the horseshoe discussed in Chapter 4. Moreover it forms the basis of the behaviour the nonuniformly hyperbolic original map H. The key idea here is to show that we have certain properties for the induced map H_S, and that these properties are shared by H. It would be possible to work entirely with H_S, but for the problem, as Przytycki (1983) points out, that checking the Katok–Strelcyn conditions for H_S is not as simple as it is for H.

7

Ergodicity and the Bernoulli property for toral linked twist maps

In this chapter we apply a global geometric argument to extend the result of the previous chapter to ergodicity and the Bernoulli property for toral linked twist maps. Conditions are given such that these results hold.

7.1 Introduction

As discussed in Chapter 3, the property of ergodicity is a long way down in the ergodic hierarchy. For the strongest mixing behaviour, we would like our linked twist maps to possess the Bernoulli property. Fortunately, the Katok–Strelcyn version of Pesin theory given in Chapter 5 gives conditions to show exactly that. Recall that if the Katok–Strelcyn conditions are satisfied, and Lyapunov exponents are non-zero for every tangent vector, and for almost every point, we have the existence of local stable and unstable manifolds for almost all points in our domain for a smooth dynamical system with singularities. Furthermore, if some forward iterate of the local unstable manifold of some point intersects some backward iterate of the local stable manifold of another point, for almost every pair of points (the Manifold Intersection Property), then the ergodic partition we showed to exist in the previous chapter has only one component, and so our linked twist map is ergodic. Moreover, if every (far enough) forward iterate of the local unstable manifold intersects some (far enough) backward iterate of the local stable manifold, again for almost every pair of points (the Repeated Manifold Intersection Property), then the linked twist map has the Bernoulli property.

In this chapter we prove that both these conditions hold for toral linked twist maps, following the work of Wojtkowski (1980) and, mainly, Przytycki (1983). We will be concerned with manipulating pieces of local stable or unstable manifold. We will, with a justification given, work exclusively with linked twist maps with linear twist functions, in which case local stable and unstable manifolds form straight lines, and we will refer to portions of them as line segments. The geometrical argument is based on the fact that line segments contained in local unstable manifolds grow in length under forward iteration, while line segments contained in local stable manifolds grow under backward iteration.

As usual, the theory for linked twist maps is anticipated by the simple case of the Arnold Cat Map. Here global stable manifolds are infinitely long straight lines and have the same slope at each point (because this system is linear and uniformly hyperbolic). Hence line segments and iterates of them are also straight lines with the same slope (recall Figures 3.1 and 5.3). Because they go in transverse directions, an intersection is inevitable. This is a straightforward result to show, but in this chapter we give rather more details than is necessary, in order to elucidate the more involved method of proof for the linked twist maps.

The case of toral linked twist maps is more complicated for three main reasons. First, iterates of line segments may straddle the boundaries of the annuli, and so contain sharp angles. The idea behind the proof is to show that such sharp bends do not prevent an intersection from occurring. As in the previous chapter, the co-rotating case is relatively straightforward, while the counter-rotating case is more complicated, and requires extra conditions. Second, the nonuniformity of the hyperbolicity means that local stable and unstable manifolds, and therefore line segments may have different slopes for different points (recall for example Figure 6.9). It is possible to compute the slopes of stable and unstable manifolds for each point explicitly, as we describe shortly, but we will only need to use the fact that stable and unstable manifolds at all points (for which they are defined) are aligned within their appropriate cones in tangent space. Third, if the twist functions are not linear functions, the iterated line segments will not be straight lines, but may be curved (recall Figure 6.7). Working with linear functions makes the notation and geometric argument simpler, and we give the justification in Section 7.2.2 for why these arguments also hold in the nonlinear case.

Until Section 7.5 the ideas in this chapter are fundamentally very simple, and are simply based on exponential stretching of line segments. This concept will be very familiar to many readers, in the guise of stretching of fluid elements. This has been a subject of interest in fluid mixing in its own right for many years (Ottino (1989a)), and is currently a source of research from the topological point of view. See for example Boyland *et al.* (2000), Thiffeault (2005), Thiffeault (2004).

7.2 Properties of line segments

We begin by making the same simplifications as in Chapter 6 which will serve to make the following arguments notionally, and notationally, more straightforward. First, we assume that the twist functions f and g are linear. Second, we assume that the strengths of the twists are equal, i.e., $|\alpha| = |\beta|$. Third, we take $\alpha > 0$, so that for the co-rotating case we have $\alpha > 0$, $\beta = \alpha > 0$, and for the counter-rotating case we have $\alpha > 0$, $\beta = -\alpha < 0$.

Throughout this chapter we shall be concerned with the behaviour of iterates of local stable and unstable manifolds. We concentrate on the behaviour of unstable manifolds γ^u under forward iteration, and note that all conclusions will also hold for the behaviour of stable manifolds γ^s under backward iteration. In the linear case, a forward iterate of γ^u, that is, $H^i(\gamma^u)$, forms a *segmented line*, that is, a union of line segments. In the following section we clarify this idea.

7.2.1 Definition, iteration and orientation of line segments

Figure 7.1 shows a sketch of a line segment γ centred at a point $z = (x, y)$. We will need to measure the horizontal and vertical lengths of line segments, and so for any line segment γ we define the following (Figure 7.1 also illustrates the vertical and horizontal lengths of the local stable manifold at a point z).

Definition 7.2.1 ($l_h(\gamma)$) *For a line segment γ beginning at (x_b, y_b) and ending at (x_e, y_e) we define the horizontal length $l_h(\gamma)$ to be*

$$l_h(\gamma) = x_e - x_b.$$

Definition 7.2.2 ($l_v(\gamma)$) *For a line segment γ beginning at (x_b, y_b) and ending at (x_e, y_e) we define the vertical length $l_v(\gamma)$ to be*

$$l_v(\gamma) = y_e - y_b.$$

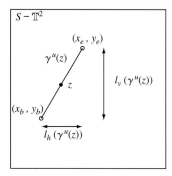

Figure 7.1 The line segment $\gamma^s(z)$, with ends (x_b, y_b) and (x_e, y_e) marked with empty circles, and with horizontal and vertical lengths labelled with arrows.

All of our line segments shall be oriented so that $y_e > y_b$, that is, so that $l_v(\gamma)$ is always positive. However $l_h(\gamma)$ may be positive or negative, depending on the slope of γ.

We note here a technical point. Strictly speaking, in order to measure the lengths of line segments which may wrap around the torus, we should do so in terms of a *lift*. This is a map from the torus $\mathbb{T}^2$ to the real plane $\mathbb{R}^2$ and can be thought of as the opposite of a *projection*. Thus given a map $f : \mathbb{T}^2 \to \mathbb{T}^2$ and a projection $\pi : \mathbb{R}^2 \to \mathbb{T}^2$, the lift of f is a (non-unique) map $F : \mathbb{R}^2 \to \mathbb{R}^2$ defined by $\pi \circ F = f \circ \pi$, so that $f(x \pmod 1) = F(x) \pmod 1$. However, we will ignore this technicality for three reasons. First, the horizontal and vertical lengths of line segments are sufficiently intuitive concepts that introducing extra notation would be unnecessary, second, we will be interested in the growth of line segments only *until* they begin to wrap more than once around the torus, and third, because working on the torus itself allows us to refer back to earlier figures more directly. However, it should be noted that many ideas in this chapter are more correctly formalized via a lift.

Arnold Cat Map

We illustrate the underlying idea with the Arnold Cat Map. The very special structure (i.e., the linearity and the uniform hyperbolicity) means that the Cat Map has local stable and unstable manifolds which are infinitely long lines, wrapping around the torus, and filling it densely (see Example 5.2.1). This of course gives us all the structure we need immediately, but to make a connection with toral linked twist maps, and with stretching of individual fluid elements, we note that any finite-sized line segment γ contained in $\gamma^u(z)$ (resp. $\gamma^s(z)$) is a segment aligned in the direction of the unstable (resp. stable) eigenvector. That is, writing $u = l_h(\gamma)$ and $v = l_v(\gamma)$ we have

$$\gamma \in \{v/u = (\sqrt{5} + 1)/2\} \text{ for all } \gamma \subset \gamma^u(z)$$

$$\gamma \in \{v/u = (1 - \sqrt{5})/2\} \text{ for all } \gamma \subset \gamma^s(z)$$

for every z.

Toral linked twist maps

For the toral linked twist map things are slightly more complicated. Here local manifolds only exist for almost every point, and they do not all share a common direction. In Chapter 6 we used the idea of first return maps F_S, G_S and H_S, and observed that the fact that the return maps may be different at different initial points produced the nonuniformity in the hyperbolicity. Recall from Section 6.2.3 that since the definitions of the return maps $F_S(z) = F^m(z)$ and

$G_S = F^{m'}(z)$ produce integers $m = m(z)$ and $m' = m'(z)$, iterating along an orbit gives a sequence of integers m_i and m'_i, resulting from:

$$H^n_S(z) = G^{m'_{n-1}} F^{m_{n-1}} G^{m'_{n-2}} F^{m_{n-2}} \cdots G^{m'_1} F^{m_1} G^{m'_0} F^{m_0}(z)$$

where each m_i, m'_i depends on the value of z along the orbit.

We note here, for completeness, that in the case of linear twists, Wojtkowski (1980) uses just such a sequence to construct local stable manifolds explicitly. For almost every $z \in \mathbb{T}^2$ define the continued fraction

$$\mathcal{H}(z) = \cfrac{1}{-\alpha m_0 + \cfrac{1}{-\beta m'_0 + \cfrac{1}{-\alpha m_1 + \cfrac{1}{-\beta m'_1 + \cdots}}}}$$

The value of this continued fraction, which clearly depends on the point z, gives the slope of the local stable manifold at each z. The slope of the local unstable manifold at each z can similarly be computed precisely. See Wojtkowski (1980) for more details. We simply observe that although local unstable manifolds may lie in different directions at each point, these directions are restricted. There is a crucial difference here between the co-rotating and counter-rotating systems. Consider a point $z = (x, y)$ to be a point for which $\gamma^u(z)$ exists, and consider $H^i(\gamma^u(z))$. This is typically a segmented line (see Figure 7.2, or recall Figures 6.9 and 6.11), whose components have slopes given by the invariant cones discussed in Chapter 6.

Co-twisting case Recall from Chapter 6 that each vector $v = (v_1, v_2)$ in tangent space with v_1, $v_2 > 0$ gave rise to a positive Lyapunov exponent, and

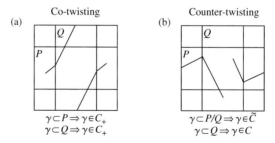

Figure 7.2 (a) For a co-twisting toral linked twist map, all segments in the segmented line $H^i(\gamma^u(z))$ lie in the cone C_+. (b) By contrast, in the counter-twisting case, $\gamma \in C$ if $\gamma \subset Q$, and $\gamma \in \tilde{C}$ if $\gamma \subset P$.

similarly for vectors with $v_1, v_2 < 0$. The same reasoning reveals that for any z such that a local unstable manifold $\gamma^u(z)$ exists, it lies in the sector $C_+ = \{(u, v) : u/v \geq 0\}$. Moreover any component γ of the segmented line $H^i(\gamma^u(z))$ also lies in this sector. That is, again by writing $u = l_v(\gamma), v = l_h(\gamma)$,

$$\gamma \in C_+ = \{u/v \geq 0\} \text{ for all } \gamma \subset H_S^i(\gamma^u(z)), \text{ for all } i \geq 0.$$

This is illustrated in Figure 7.2(a).

Counter-twisting case The situation in the counter-twisting case is slightly more complicated. Recall from Chapter 6 and Figure 6.10 the definition of the cones C and $\tilde{C}$:

$$C(z) = \{(u, v) : L \leq u/v \leq 0\}$$

$$\tilde{C}(z) = \{(u, v) : u/v \geq L + \alpha\}$$

where $L = -\alpha/2 + \sqrt{(\alpha/2)^2 - 1}$. These were a pair of cones such that $DF(C) \subseteq \tilde{C}$ and $DG(\tilde{C}) \subseteq C$. Again the same argument reveals that for a component γ in the segmented line $H^i(\gamma^u(z))$ lying in Q (so that the map G has just been applied), γ lies in the cone C. Conversely, for a segment γ which lies outside Q (i.e., in $P \backslash Q$, so that F was the last map to be applied), γ lies in $\tilde{C}$. This is illustrated in Figure 7.2(b), and we can summarize this as:

$$\gamma \in C = \{L \leq u/v \leq 0\} \text{ for all } \gamma \subset H_S^i(\gamma^u(z)) \cap Q, \text{ for all } i \geq 0$$

$$\gamma \in \tilde{C} = \{u/v \geq L + \alpha\} \text{ for all } \gamma \subset H_S^i(\gamma^u(z)) \cap P \backslash Q, \text{ for all } i \geq 0$$

where, as in Chapter 6, $L = (-\alpha/2) + \sqrt{(\alpha/2)^2 - 1}$.

Key point: A co-rotating toral linked twist map is co-twisting, which means that segments of segmented lines all lie in roughly the same direction. Conversely, a counter-rotating toral linked twist map is counter-twisting, and acute angles can form between line segments.

With a line segment $\gamma(z)$ defined as above we define for clarity the image under the first return map.

Definition 7.2.3 ($F_S(\gamma(z))$) *Given a line segment $\gamma(z)$ the image under the first return map F_S is*

$$F_S(\gamma(z)) = F^m(\gamma(z))$$

where m is such that $F^m(z) \in S$, $F^i(z) \notin S$ for $0 \leq i < m$.

Thus the first return of a line segment is given by the first return of that line segment's centre point, and not by the first return of any part of the segment. We shall often refer to $F_S(\gamma(z))$ as $F_S(\gamma)$ for simplicity of notation, and similarly refer to $\gamma(z)$ as γ. Similarly we can define:

Definition 7.2.4 ($G_S(\gamma(z))$) *Given a line segment $\gamma(z)$ the image under the first return map G_S is*

$$G_S(\gamma(z)) = G^{m'}(\gamma(z))$$

where m' is such that $G^{m'}(z) \in S$, $G^i(z) \notin S$ for $0 \le i < m'$.

Definition 7.2.5 ($H_S(\gamma(z))$) *Given a line segment $\gamma(z)$ the image under the first return map H_S is*

$$H_S(\gamma(z)) = H^{m''}(\gamma(z))$$

where m'' is such that $H^{m''}(z) \in S$, $H^i(z) \notin S$ for $0 \le i < m''$.

7.2.2 Growth of line segments

A very simple but key result which we shall refer to repeatedly throughout this chapter is the growth of a line segment due to a single twist map. A horizontal twist map F fixes the vertical length of a line segment γ and alters the horizontal length by an amount depending on the vertical length, whilst a vertical twist map G fixes the horizontal length of a line segment γ and alters the vertical length by an amount depending on the horizontal length. In particular we have the following obvious facts:

$$l_v(F(\gamma)) = l_v(\gamma) \tag{7.1}$$

$$l_h(G(\gamma)) = l_h(\gamma) \tag{7.2}$$

and the following straightforward lemmas:

Lemma 7.2.1 *For $\gamma \in C_+ = \{u/v \ge 0\}$,*

$$l_h(F(\gamma)) = l_h(\gamma) + \alpha l_v(\gamma).$$

Proof Labelling the end points of γ as (x_b, y_b), (x_e, y_e) as in Figure 7.1 we have

$$
\begin{aligned}
l_h(F(\gamma)) &= (x_e + f(y_e)) - (x_b + f(y_b)) \\
&= (x_e - x_b) + (f(y_e) - f(y_b)) \\
&= l_h(\gamma) + \alpha(y_e - y_b) \\
&= l_h(\gamma) + \alpha l_v(\gamma)
\end{aligned}
$$

since for a linear function $f(y)$, we have $\alpha = f'(y) = (f(y_e) - f(y_b))/(y_e - y_b)$ for all y_b, y_e. ☐

Lemma 7.2.2 *For $\gamma \in C_+ = \{u/v \geq 0\}$,*

$$l_v(G(\gamma)) = l_v(\gamma) + \alpha l_h(\gamma).$$

Proof Labelling the end points of γ as before we have

$$
\begin{aligned}
l_v(G(\gamma)) &= (y_e + f(x_e)) - (y_b + f(x_b)) \\
&= (y_e - y_b) + (f(x_e) - f(x_b)) \\
&= l_v(\gamma) + \alpha(l_h(\gamma)).
\end{aligned}
$$

☐

For the case of the Cat Map, the growth of line segments occurs at every iteration, while for a linked twist map, the growth of line segments only occurs for the portions of line segments which lie inside the relevant annulus. Note that the above lemmas apply only for line segments $\gamma \in C_+$; that is, they only apply in the co-rotating case. We will need a different growth criterion for the counter-rotating case, to be discussed in Section 7.5.

The nonlinear case If the twist functions f and g are nonlinear functions (defined as before with strength α as the minimum derivative), all the results in this chapter hold, as we can replace every occurence of the above growth lemmas with, for example

$$
\begin{aligned}
l_h(F(\gamma)) &= (x_e + f(y_e)) - (x_b + f(y_b)) \\
&= l_h(\gamma) + f(y_e) - f(y_b) \\
&= l_h(\gamma) + f'(y)(y_e - y_b)
\end{aligned}
$$

for some $y \in (y_b, y_e)$, by the Mean Value Theorem. Since $f'(y) \geq \alpha$ by definition, we have

$$l_h(F(\gamma)) \geq l_h(\gamma) + \alpha l_v(\gamma).$$

7.2.3 *v*-segments and *h*-segments

We are interested in proving that an intersection of $H^m(\gamma^u(z))$ and $H^{-n}(\gamma^s(z'))$ must occur. Such an intersection could occur in many ways, but one way to guarantee it is to find h- and v-segments, which are defined as follows:

Definition 7.2.6 (h-segment) *A line segment in S joining the left and right boundaries of S (that is, a line segment γ such that $\gamma \subset S$, $\gamma \cap \partial Q_0 \neq \emptyset$ and $\gamma \cap \partial Q_1 \neq \emptyset$) is called an h-segment and is denoted γ_h.*

Definition 7.2.7 (v-segment) *A line segment in S joining the top and bottom boundaries of S (that is, a line segment γ such that $\gamma \subset S$, $\gamma \cap \partial P_0 \neq \emptyset$ and $\gamma \cap \partial P_1 \neq \emptyset$) is called a v-segment and is denoted γ_v.*

Note that any h-segment must intersect any v-segment. Figure 7.3(a) shows a v-segment intersecting an h-segment. Thus if we can show that $H^m(\gamma^u(z))$ contains an h-segment, and $H^{-n}(\gamma^s(z'))$ contains a v-segment, then an intersection must occur and we have satisfied the Manifold Intersection Property.

We note here an issue with working on the torus. Figure 7.4 shows a line segment γ_v which wraps vertically around the torus, connecting upper and lower sides of the square. This, by definition, is a v-segment, but it also wraps (partially) around the torus horizontally, cutting the left and right boundaries.

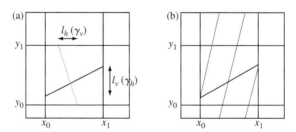

Figure 7.3 Diagram (a) shows an h-segment γ_h (solid line), and a v-segment γ_v (dotted line), together with the horizontal length l_h of γ_v and the vertical length l_v of γ_h. Diagram (b) shows an h-segment γ_h (solid line) and its image (dotted line) under an iteration of G for $l = 2$ (a double twist). Note that $G(\gamma_h)$ contains a v-segment.

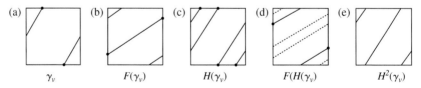

Figure 7.4 Diagram showing how after at most two iterates of the Cat Map, an initial v-segment γ_v as shown in (a) must produce a v-segment which does not cut the left and right sides of the square.

This makes it possible to construct an h-segment (a line connecting left and right sides) which does not intersect γ_v. (Alternatively, one could argue that a v-segment as in Figure 7.3 might fail to cross an h-segment which wraps around the torus vertically). We show diagramatically in Figure 7.4 that iterating any v-segment (e.g., that in Figure 7.4(a)), must lead to a v-segment which connects upper and lower sides of the square without cutting the left and right sides.

Figure 7.4(a) shows a v-segment γ_v. In this and subsequent figures, solid circles mark points fixed by the map to be applied. Figure 7.4(b) shows $F(\gamma_v)$, after an iterate of the horizontal twist. This is a line connecting lower and upper sides of the square which wraps horizontally around the torus more than once (here 'more than once' means 'some fraction more than one complete wrapping'). Figure 7.4(c) shows the image of this line under G, so that $G(F(\gamma_v)) = H(\gamma_v)$ contains a line connecting lower and upper sides of the square, wrapping twice vertically and more than once horizontally. Applying F to this line, we get the image in 7.4(d). This is a line (marked solid and dashed) connecting lower and upper sides of the square, wrapping twice vertically and more than three times horizontally. Crucially it contains the line marked in solid connecting the left and right sides which cut the lower and upper sides of the square. In 7.4(e) we show only the image of the solid line, which must necessarily contain a v-segment.

There are many similar ways to express this argument, including working with the lift of the map, or using combinatorial or algebraic arguments. They are all based on the fundamental premise that line segments for the Cat Map continue to grow indefinitely under iteration of the map. The same argument can be made for the co-rotating linked twist map, and the corresponding figure to Figure 7.4 is shown in Figure 7.5. Note however that for the linked twist map case, defining a v-segment to be a line entirely contained in S which joins ∂P_0 to ∂P_1 ensures that this issue does not arise.

We will use the same fundamental method of proof to show the Manifold Intersection Property for the Cat Map, for co-twisting toral linked twist maps,

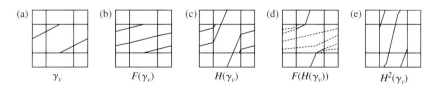

(a) γ_v (b) $F(\gamma_v)$ (c) $H(\gamma_v)$ (d) $F(H(\gamma_v))$ (e) $H^2(\gamma_v)$

Figure 7.5 Diagram showing how after at most two iterates of a co-rotating toral linked twist map, an initial v-segment γ_v as shown in (a) must produce a v-segment which does not intersect ∂Q_0 or ∂Q_1.

and then for counter-twisting toral linked twist maps. Consider a line segment $\gamma \subset \gamma^u(z)$. We aim to prove that we have a v-segment after some iterate of H. We do so by showing the exponential growth of $H^i(\gamma^u(z))$; that is, by showing that $l_v(H^i(\gamma)) \geq K^i l_v(\gamma)$ for some constant $K > 1$. This exponentially growing segment must result in a v-segment.

7.3 Ergodicity for the Arnold Cat Map

In this section we give a proof that the Arnold Cat Map satisfies the Manifold Intersection Property. This, in effect, confirms that the Cat Map is ergodic (that is, that the ergodic partition guaranteed by the fact that Lyapunov exponents are non-zero has only a single component of positive measure). Of course, this is a well-established result in its own right, and moreover, as discussed in Chapter 3, we can show topological transitivity for the Cat Map directly. We give this alternative argument as a precursor to the geometrically more complicated arguments needed for linked twist maps.

Recall that the Arnold Cat Map can be written as $H = GF$, with $F(x, y) = (x + y, y)$ and $G(x, y) = (x, x + y)$, so that $\alpha = 1$. We have already shown that for each $z \in \mathbb{T}^2$, the local unstable manifold γ^u exists. Moreover $\gamma^u \in C_+$. Then the above lemmas on growth of line segments produce the following.

Lemma 7.3.1 *Let* $\gamma \subset \gamma^u \in C_+$ *be any line segment contained in a local unstable manifold of the Arnold Cat Map H. Then*

$$l_v(H(\gamma)) = K_1 l_v(\gamma)$$

where $K_1 > 1$ is a constant independent of the choice of z and γ. Hence

$$l_v(H^i(\gamma)) = K_1^i l_v(\gamma)$$

for each $i \geq 0$.

Proof Choose a point $z = (x, y)$ and consider a connected subset γ of its unstable manifold $\gamma^u(z)$. Since $\gamma \in C_+$, we have

$$
\begin{aligned}
l_v(H(\gamma)) &= l_v(G(F(\gamma))) \\
&= l_v(F(\gamma)) + \alpha l_h(F(\gamma)) \\
&= l_v(\gamma) + \alpha(l_h(\gamma) + \alpha l_v(\gamma)) \\
&= (1 + \alpha^2) l_v(\gamma) + \alpha l_h(\gamma) \\
&\geq K_1 l_v(\gamma)
\end{aligned}
$$

where $K_1 = 1 + \alpha^2 > 1$ (in fact here we have $K_1 = 2$ since for the Cat Map, $\alpha = 1$. Hence the vertical length of line segments at least doubles at each iterate of H). Since K_1 is independent of the choice of z and γ we can apply this result repeatedly to acquire $l_v(H^i(\gamma)) = K_1^i l_v(\gamma)$. $\qquad\qquad\square$

Lemma 7.3.2 *The Arnold Cat Map satisfies the Manifold Intersection Property.*

Proof Lemma 7.3.1 shows that the vertical length of $H^i(\gamma)$ for $\gamma \in C_+$ grows (exponentially) with i. Therefore there exists an n such that $H^n(\gamma)$ contains a v-segment. Similarly, the horizontal length of $H^j(\gamma')$ for $\gamma' \in C_- = \{(u, v) : u/v \leq 0\}$ grows (exponentially) with j, and so there exists an m such that $H^{-m}(\gamma)$ contains an h-segment. Since any v-segment intersects any h-segment, the Manifold Intersection Property is satisfied. $\qquad\qquad\square$

Corollary 7.3.1 *The Arnold Cat Map is ergodic.*

7.4 Ergodicity for co-rotating toral linked twist maps

The corresponding result for co-rotating linked twist maps is notionally easy. Line segments grow in a similar manner to the Cat Map. In particular, because $\gamma \in C_+$ for each $\gamma \in H^i(\gamma^u(z))$, we have $l_v(H^{i+1}(\gamma)) \geq l_v(H^i(\gamma))$. This growth of line segments should lead to a v-segment as before, but there is an additional issue here. Recall that linked twist maps, being non-uniformly hyperbolic systems, have local unstable manifolds, the sizes of which are not uniformly bound away from zero. This means that we must ensure that the rate of growth of the vertical length of line segments is sufficiently great to ensure a v-segment in finite time, no matter how small the initial γ^u. We require exponential growth (as we had for the Cat Map) to achieve this, and this can be obtained using the result from Chapter 6 (Lemma 6.3.3) which implies that any orbits hitting S do so with positive frequency.

Lemma 7.4.1 *Let $z \in S$ be such that $\gamma^u(z)$ exists. For each $\gamma \in C_+$*

$$l_v(H_S(\gamma)) \geq K_2 l_v(\gamma)$$

where $K_2 > 1$ is a constant independent of the choice of z and γ. Hence

$$l_v(H_S^i(\gamma)) \geq K_2^i l_v(\gamma)$$

until $H_S^i(\gamma)$ contains a v-segment.

Proof Since z is the mid-point of γ, at least half of (the vertical length of) γ is acted on by F_S. That is, letting $\gamma' = F_S(\gamma)$,

$$l_h(\gamma') \geq l_h(\gamma) + \frac{\alpha}{2} l_v(\gamma).$$

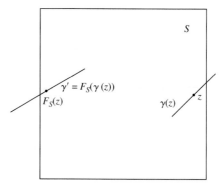

Figure 7.6 The line segments $H_S^r(\gamma(z))$ and $F_S(H_S^r(\gamma(z)))$, with chosen segments γ and γ'.

Moreover, since $l_v(\gamma') = l_v(\gamma)$, and $F_S(z) \in S$, at least half of (the horizontal length of) γ' is acted on by G_S (see Figure 7.6). That is,

$$l_v(H_S(\gamma)) = l_v(G_S(\gamma')) \geq l_v(\gamma') + \frac{\alpha}{2} l_h(\gamma')$$

$$\geq l_v(\gamma) + \frac{\alpha}{2}\left(l_h(\gamma) + \frac{\alpha}{2} l_v(\gamma)\right)$$

$$\geq K_2 l_v(\gamma)$$

where $K_2 = 1 + \alpha^2/4 > 1$. Since K_2 is independent of the choice of z and γ, we can continue this process so that $l_v(H_S^i(\gamma)) \geq K_2^i l_v(\gamma)$ until we reach a v-segment.[1]   □

Theorem 7.4.1 *A co-rotating toral linked twist map H satisfies the Manifold Intersection Property.*

Proof Lemma 7.4.1 shows that the vertical length of $H^i(\gamma)$ for $\gamma \in C_+$ grows exponentially until it contains a v-segment. The same argument can be applied to $H^{-i}(\gamma')$ for $\gamma' \in C_-$ to produce an h-segment. Since these must intersect, the Manifold Intersection Property is satisfied.   □

Corollary 7.4.1 *A co-rotating toral linked twist map H is ergodic.*

> **Key point:** The proof of the Manifold Intersection Property, and hence ergodicity, in the co-rotating/co-twisting case is fundamentally very simple. Forward iterates of unstable manifolds grow and wrap around the torus, while backward iterates of stable manifolds grow and wrap around the torus in a different direction, making an intersection inevitable.

---

[1] Alert readers will recognize that this proof may fail for a trajectory which, for example, approaches the top left corner of S. It can be shown that initial conditions producing such trajectories form a set of zero measure in S.

7.5 Ergodicity for counter-rotating toral linked twist maps

The counter-rotating case has an added difficulty, in that the counterpart to
Lemma 7.4.1 does not hold. In particular, observe that it is not necessarily true
that $l_v(H(\gamma)) > l_v(\gamma)$. This is because iterates of unstable manifolds, while
lengthening, may fold back on themselves and a priori fail to wrap around
the torus, producing h- and v-segments. Figure 7.7 provides a sketch of this, after
Przytycki (1983). Iterates of a segment of local unstable manifold γ^u increase
in length but remain confined in a region of phase space. See Figure 6.12 of
Chapter 6 for this type of behaviour in a real numerical example (although recall
that this example also violates the conditions for non-uniform hyperbolicity).
These difficulties were overcome by Przytycki (1983), who gave an intricate and
ingenious argument to deduce exponential growth of line segments culminating
in an h-segment. We reproduce his proof.

First we note that the following arguments will require at least double twists;
that is, the wrapping numbers k and l will have modulus at least 2. This
immediately allows the following lemma:

Lemma 7.5.1 *For a counter-rotating toral linked twist map with $|l| \geq 2$, let
γ be a line segment. If $F(\gamma)$ contains an h-segment γ_h, then $H(\gamma)$ contains a
v-segment.*

Proof Since $|l| > 2$ the map G wraps γ_h at least twice around the torus, with
the end points of γ_h fixed, as shown in Figure 7.3(b). Hence $H(\gamma)$ must contain
a v-segment. $\qquad\square$

We aim to show that, given $\gamma \subset H_S^i(\gamma^u(z))$ (i.e., a line segment in S aligned
in the cone C), **either** $F_S(\gamma)$ contains a segment γ' such that

$$l_h(\gamma') > \delta l_v(\gamma) \tag{7.3}$$

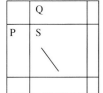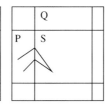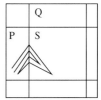

Figure 7.7 Schematic diagram (after Przytycki (1983)) showing how iterates of
γ^u in a counter-rotating map may grow but fail to form h- or v-segments. The
first diagram shows an initial line segment γ, and subsequent diagrams show the
images of γ under subsequent iterates of H.

where $\delta > 1$ is independent of γ, **or** $F_S(\gamma)$ contains an h-segment. In the latter case, we can appeal immediately to Lemma 7.5.1 to produce a v-segment. In the former case we repeat the process, acting on the line segment γ' with G_S, to find a segment $\gamma'' \subset G_S(\gamma')$ such that

$$l_v(\gamma'') > \delta l_h(\gamma')$$

unless γ'' is already a v-segment. We continue this process to produce line segments of exponentially growing length, which must eventually end with a v-segment or an h-segment. To establish Equation 7.3 we first make the following assumption.

Assumption $F_S(\gamma)$ does not contain any h-segments.

Clearly if the assumption does not hold then we have our h-segment and can appeal immediately to Lemma 7.5.1.

Lemma 7.5.2 *A consequence of the assumption is that* $l_v(\gamma)(L + \alpha) < 2$.

Proof We have $\gamma \in C = \{(u, v) : L \leq u/v \leq 0\}$, so $Ll_v(\gamma) \leq l_h(\gamma) \leq 0$. To guarantee that $F_S(\gamma)$ may not contain an h-segment we must have $\frac{1}{2}l_h(F_S(\gamma)) < 1$ (this is because if $\frac{1}{2}l_h(F_S(\gamma)) \geq 1$ then even if the 'worst case' of $S = \mathbb{T}^2$ and $F_S(z)$ close to ∂Q we must have an h-segment). But $l_h(F_S(\gamma)) \geq l_h(\gamma) + \alpha l_v(\gamma) \geq (L + \alpha)l_v(\gamma)$. So a consequence of the assumption is that $(L + \alpha)l_v(\gamma) < 2$.   □

Let m_1 be the first time $F^{m_1}(\gamma)$ intersects S. There are four cases in which such an intersection can occur:

1. $F^{m_1}(\gamma)$ contains an h-segment γ_h.
2. The right-hand end of $F^{m_1}(\gamma)$ intersects S (see Figure 7.8).
3. The left-hand end of $F^{m_1}(\gamma)$ intersects S.
4. Both ends of $F^{m_1}(\gamma)$ intersect S (see Figure 7.8).

In case (1) we may apply Lemma 7.5.1 on γ_h immediately. Cases (2) and (3) are analogous. We investigate first (2). Divide $F^{m_1}(\gamma)$ into intervals I_1, I_2, I_3, I_4, as in Figure 7.8. We make the division so that $F^{m_1}(\gamma) \cap S = I_4$, but the remainder of the division into I_1, I_2, I_3 is arbitrary.

Lemma 7.5.3 *There exists a periodic point p in I_2.*

Proof Along I_2, $f(y)$ changes by $\alpha l_v(I_2)$. (To see this label the end points of I_2 as (I_{x_s}, I_{y_s}) and (I_{x_e}, I_{y_e}) and note the linearity of the twist gives $f(I_{y_e}) - f(I_{y_s}) = \alpha(I_{y_e} - I_{y_s}) = \alpha l_v(I_2)$.) Since f is continuous we can choose $n > 1/\alpha l_v(I_2)$ such that there exists $y \in (I_{y_s}, I_{y_e})$ with $f(y)$ a rational number m/n for some

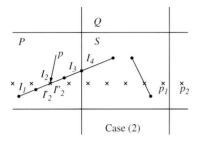
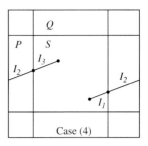

Figure 7.8 (Figure after Przytycki (1983)) Case (2) shows the initial line segment γ and the image $F^{m_1}(\gamma)$. We partition $F^{m_1}(\gamma)$ as shown into I_1, I_2, I_3, I_4. The segment I_2 contains a periodic point p whose orbit is marked with crosses. Case (4) has the image $F^{m_1}(\gamma)$ intersecting S from the left and right.

integer m. This implies that there is an F-periodic point $p \in I_2$ with period $n = \lfloor 1 + 1/(\alpha l_v(I_2)) \rfloor$. The distance between points in this periodic orbit is

$$d = 1/n \geq 1/(1 + (1/\alpha l_v(I_2))). \tag{7.4}$$

$\square$

The point p divides I_2 into two parts which we call I_2' and I_2''. Of the points in the (periodic) orbit of p we label the last one to the left of ∂Q_1 as p_1, and the first to the right of ∂Q_1 as p_2. Now call m_2 the first time that $F^{m_2}(p)$ is between p and ∂Q_1. (Note that there must be such a point as the orbit of p a point in $\gamma \in S$. The point could be p_1, but needn't be.) Let us now inspect $I_2'' \cup I_3$. Define

$$J_0 = I_2'' \cup I_3$$
$$J_m = F(J_{m-1} \backslash S) \quad m = 1, 2, \ldots, m_2.$$

We define the sequence in this way, discarding any portions which intersect S before applying F so that we only ever act on J_i with F, and never G. This means we never introduce acute angles via the counter-rotation. However the final element in the sequence, J_{m_2} may intersect S. We shall see shortly that in fact it must. Similarly define

$$\bar{J}_0 = I_1 \cup I_2'$$
$$\bar{J}_m = F(\bar{J}_{m-1} \backslash S) \quad m = 1, 2, \ldots, m_2.$$

Then we study the growth of $(J_{m_2} \cup \bar{J}_{m_2}) \cap S$, since this is equivalent to the growth of portions of $I_1 \cap I_2 \cap I_3$ which have only been acted on by one twist map, F, which finally intersect S. It is enough if this quantity satisfies the growth criteria of Equation (7.3).

Lemma 7.5.4 *To have $l_h((J_{m_2} \cup \bar{J}_{m_2}) \cap S) \geq \delta l_v(\gamma)$ we need all of the following inequalities to hold:*

$$d \geq \delta l_v(\gamma), \tag{7.5}$$

$$\alpha l_v(I_3) + l_h(I_3) \geq \delta l_v(\gamma), \tag{7.6}$$

$$\alpha l_v(I_1) + l_h(I_1) \geq \delta l_v(\gamma). \tag{7.7}$$

Proof First we note that for $m = 1, 2, \ldots, m_2$,

$$l_h(J_m) \geq \min\{l_h(I_2'' \cup I_3) + \alpha l_v(I_2'' \cup I_3), d + l_h(I_2'' \cup I_3)\}.$$

To see this recall that m_2 is the first time that $F^{m_2}(p)$ is between p and ∂Q_1. If J_m intersects ∂Q_0 then $l_h(J_m)$ is at least $d + l_h(I_2'' \cup I_3)$. If J_m does not intersect S, then the usual growth of a line segment applies and $l_h(J_m)$ is $l_h(I_2'' \cup I_3) + \alpha l_v(I_2'' \cup I_3)$. Now we examine $l_h(J_{m_2} \cap S)$. If $F^{m_2}(p)$ is between p and ∂Q_0 then $l_h(J_{m_2} \cap S) \geq d$. If $F^{m_2}(p) \in S \setminus \{p_1\}$ then

$$l_h(J_{m_2} \cap S) \geq \min\{d, l_h(I_2'' \cup I_3) + \alpha l_v(I_2'' \cup I_3)\}. \tag{7.8}$$

Finally we have the case $F^{m_2}(p) = p_1$. If Equation (7.8) is not satisfied then $J_{m_2} \cap S$ intersects ∂Q_1 with its right-hand end and we have $l_h(J_{m_2} \cap S) = \tau d$ for some $0 < \tau < 1$.

A similar argument applies for $\bar{J}$ giving

$$l_h(\bar{J}_{m_2} \cap S) \geq \min\{(1 - \tau)d, l_h(I_1 \cup I_2') + \alpha l_v(I_1 \cup I_2')\}. \tag{7.9}$$

unless $\bar{J}_{m_2} \cap S$ intersects ∂Q_0 with its left-hand end.

This gives

$$l_h((J_{m_2} \cup \bar{J}_{m_2}) \cap S) > \min\{d, l_h(I_3) + \alpha l_v(I_3), l_h(I_1) + \alpha l_v(I_1)\}. \tag{7.10}$$

$\square$

We would also have Equation (7.3) satisfied if

$$l_h(I_4) \geq \delta l_v(\gamma). \tag{7.11}$$

Lemma 7.5.5 *There exists a partition, given below, of γ into I_1, I_2, I_3, I_4 satisfying all of Equations (7.5), (7.6) and (7.7), or Equation (7.11), if $\alpha > \max\{\alpha_1, \alpha_2\} = \alpha_2$, where $\alpha_1 = 5/2$ and $\alpha_2 \approx 4.152643$.*

Proof From the definition of d (Equation (7.4)), Equation (7.5) is equivalent to

$$\frac{\alpha l_v(I_2)}{1 + \alpha l_v(I_2)} \geq \delta l_v(\gamma).$$

We can rearrange this to give

$$l_v(I_2) \geq \frac{\delta l_v(\gamma)}{\alpha(1 - \delta l_v(\gamma))}. \tag{7.12}$$

This holds since $1 - \delta l_v(\gamma) > 0$. To see this note that $l_v(\gamma) < 2/(L+\alpha)$ from Lemma 7.5.2, and $L + \alpha = (\alpha/2) + \sqrt{(\alpha/2)^2 - 1} > 2$ for $\alpha > \alpha_1 = 5/2$. Equation (7.6) is equivalent to

$$l_v(I_3) \geq \frac{\delta l_v(\gamma)}{L + 2\alpha}. \tag{7.13}$$

This follows from the facts that $\gamma \in C = \{(u, v) : L \leq u/v \leq 0\}$ and so $l_h(I_3) \geq (L + \alpha)l_v(I_3)$. We have a similar conclusion for $l_v(I_1)$.

$$l_v(I_1) \geq \frac{\delta l_v(\gamma)}{L + 2\alpha}. \tag{7.14}$$

Equation (7.11) follows if

$$l_v(I_4) \geq \frac{\delta l_v(\gamma)}{L + \alpha} \tag{7.15}$$

again using $l_h(I_4) \geq (L + \alpha)l_v(I_3)$.

Thus we need a partition of γ into I_1, I_2, I_3, I_4 such that either all of (7.12), (7.13) and (7.14) hold, or (7.15) holds. Suppose

$$l_v(\gamma) = l_v(I_1) + l_v(I_2) + l_v(I_3) + l_v(I_4)$$
$$> \frac{\delta l_v(\gamma)}{L + 2\alpha} + \frac{\delta l_v(\gamma)}{\alpha(1 - \delta l_v(\gamma))} + \frac{\delta l_v(\gamma)}{L + 2\alpha} + \frac{\delta l_v(\gamma)}{L + \alpha}. \tag{7.16}$$

Then we choose I_1 such that

$$l_v(I_1) \geq \frac{\delta l_v(\gamma)}{L + 2\alpha}$$

(which we can since $\delta < 1$ and $L + 2\alpha > \alpha_1 + \alpha > 1$) and choose I_3 such that

$$l_v(I_3) \geq \frac{\delta l_v(\gamma)}{L + 2\alpha}$$

(again this can be done). Then Equation (7.16) implies that

$$l_v(I_2) + l_v(I_4) > \frac{\delta l_v(\gamma)}{\alpha(1 - \delta l_v(\gamma))} + \frac{\delta l_v(\gamma)}{L + \alpha}$$

which means that either

$$l_v(I_2) > \frac{\delta l_v(\gamma)}{\alpha(1 - \delta l_v(\gamma))}$$

in which case we have all of (7.12), (7.13), (7.14) or

$$l_v(I_4) > \frac{\delta l_v(\gamma)}{L + \alpha}$$

so (7.15) holds. Dividing (7.16) through by $l_v(\gamma)$ gives the condition

$$1 > \frac{1}{\alpha(1 - 2/(L + \alpha))} + \frac{2}{2\alpha + L} + \frac{1}{\alpha + L}$$

which is satisfied by

$$\alpha > \alpha_2 \approx 4.152.$$

□

Case (3) follows in exactly the same way. We now study case (4). This is a slightly simpler arrangement.

Lemma 7.5.6 *There exists a partition which gives $l_h(F_S(\gamma)) > \delta l_v(\gamma)$ if $\alpha > \alpha_3 \approx 3.239$.*

Proof We divide $F^{m_1}(\gamma)$ into I_1, I_2, I_3 as in Figure 7.8. Clearly we would be done if $l_h(I_1) \geq \delta l_v(\gamma)$, or if $l_h(I_3) \geq \delta l_v(\gamma)$. Moreover, it would be enough if $l_h(F(I_2)) - l_h(I_2) \geq \delta l_v(\gamma)$. As above these would be satisfied if any one of the following inequalities hold:

$$(L + \alpha)l_v(I_1) \geq \delta l_v(\gamma),$$
$$(L + \alpha)l_v(I_3) \geq \delta l_v(\gamma),$$
$$\alpha l_v(I_2) \geq \delta l_v(\gamma).$$

This is satisfied by the partition

$$l_v(\gamma) = \sum_{i=1}^{i=3} l_v(I_i) \geq l_v(\gamma) \left(\frac{2}{L + \alpha} + \frac{1}{\alpha} \right).$$

Dividing through by $l_v(\gamma)$ gives the condition

$$1 > \frac{2}{L + \alpha} + \frac{1}{\alpha}$$

which is satisfied by

$$\alpha > \alpha_3 \approx 3.239.$$

□

Theorem 7.5.1 *A counter-rotating toral linked twist map H with $|k|$, $|l| \geq 2$ and $\alpha\beta < C = \alpha_2^2 \approx -17.24445$ is ergodic.*

Proof We proved in Chapter 6 that H satisfies the (**KS**) and (**OS**) conditions, and have positive Lyapunov exponents if $|\alpha|, |\beta| > 2$. Lemmas 7.5.5 and 7.5.6 show that if $\alpha > \alpha_2 \approx 4.152$ then Equation (7.3) is satisfied. Since $|l| \geq 2$ this guarantees a v-segment in $H^m(\gamma^u(z))$ for some $m > 0$, for almost every z. We can repeat this argument for backward iterations to conclude that there exists an h-segment in $H^{-n}(\gamma^s(z))$ for some $n > 0$, for almost every z, and so the Manifold Intersection Property is fulfilled. Then Theorem 5.4.1 proves that H is ergodic. $\qquad\square$

> **Key point:** The counter-rotating case requires a further condition on the strength of twists to guarantee exponential growth of line segments.

7.6 The Bernoulli property for toral linked twist maps

Having established ergodicity for toral linked twist maps by verifying the Manifold Intersection Property, our final task is to demonstrate that these systems possess the Bernoulli property. Recall from Section 5.4.1 that in order to show this, we must verify the Repeated Manifold Intersection Property, which is that for almost every pair of z, z',

$$H^m(\gamma^u(z)) \cap H^{-n}(\gamma^s(z)) \neq \emptyset$$

for *all* large enough integers m and n.

As usual we illustrate the argument by giving a proof that the Arnold Cat Map has the Bernoulli property. Again, this can be proved more directly, using for example, Markov Partitions (Adler (1998)). The proof of this is given with reference to Figure 7.9. Note that an alternative proof could be given by working with the lift of the Cat Map, but we choose to work on the torus in order to make the connection with figures such as those in Chapter 6. The proof of the following is based on the earlier argument about v-segments.

Theorem 7.6.1 *The Arnold Cat Map has the Bernoulli property.*

Proof We have shown that after some iterate H^i of an initial γ, we have a v-segment; that is, a line segment joining the upper and lower sides of the unit square. We consider the next few iterates of γ_v in Figure 7.9(a), recalling that the Cat Map can be regarded as the composition of a horizontal shear F and a vertical shear G. In each figure, solid black circles indicate points which are fixed under the map to be applied. Consider the initial v-segment γ_v. After F is applied

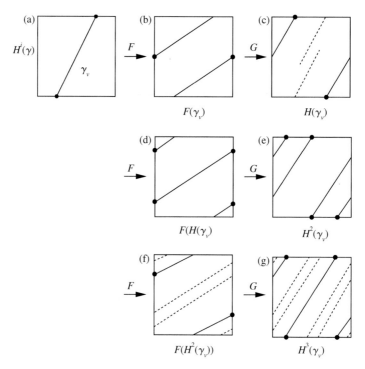

Figure 7.9 Schematic diagram illustrating how a v-segment γ_v for the Arnold Cat Map results in another v-segment after at most three iterations of H. Moreover, after three iterations, the image of γ_v has wrapped around the torus sufficiently that a v-segment is guaranteed for all future iterations.

(Figure 7.9(b)) , we have a line $F(\gamma_v)$ joining top and bottom sides of the square which wraps horizontally around the torus. After G is applied (Figure 7.9(c)) the image $H(\gamma_v)$ includes a segment which joins top and bottom sides, and cuts the left and right sides of the square. Note that here we disregard the dotted parts of the image. Applying F to the solid line segment (Figure 7.9(d)) produces a line $F(H(\gamma_v))$ connecting lower and upper sides of the square which wraps horizontally more than once. Applying G to this (Figure 7.9(e)) gives a line $H^2(\gamma_v)$ connecting lower and upper sides of the square wrapping twice vertically and more than once horizontally. Recalling our earlier argument, applying F to this line (Figure 7.9(f)) produces a line $F(H^2(\gamma_v))$ connecting lower and upper sides, wrapping more than three times horizontally, and twice vertically. In Figure 7.9(f) we have drawn part of $F(H^2(\gamma_v))$ with a solid line. After applying G the image of this line (Figure 7.9(g)) creates a v-segment which does not cut the left and right sides of the square. Note however, that

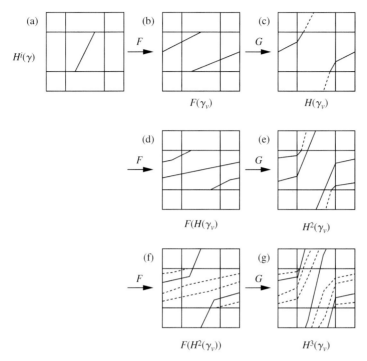

Figure 7.10 Schematic diagram illustrating how a v-segment γ_v for a co-rotating toral linked twist map results in another v-segment after at most three iterations of H. Moreover, after three iterations, the image of γ_v has wrapped around the torus sufficiently that a v-segment is guaranteed for all future iterations.

we can now apply the same argument to $H^3(\gamma_v)$, and guarantee that $H^i(\gamma_v)$ contains a v-segment for all $i \geq 3$.

As usual the same argument applies to backward iterations of local stable manifolds, to produce an h-segment for $H^{-j}(\gamma_h)$ for all $j \geq 3$. This verifies the Repeated Manifold Intersection Property. Therefore by Theorem 5.4.1 the Arnold Cat Map has the Bernoulli property.   □

Theorem 7.6.2 *A co-rotating toral linked twist map H has the Bernoulli property.*

Proof Precisely the same argument as the above applies to iterates of a v-segment γ_v of a co-rotating toral linked twist map. Figure 7.10 shows this diagramatically.   □

In the proof of ergodicity in the counter-rotating case we required the condition $|k|, |l| \geq 2$, that is, we needed double twists. This condition makes it easy to deduce the Bernoulli property for counter-rotating linked twist maps with double twists.

Theorem 7.6.3 (Przytycki (1983)) *A counter-rotating toral linked twist map H with $|k|$, $|l| \geq 2$ and $\alpha\beta < C \approx -17.24445$ has the Bernoulli property.*

Proof Theorem 7.5.1 shows that $\gamma_v \subset H^m(\gamma^u(z))$ is a v-segment. Because $|k| \geq 2$, $F(\gamma_v)$ contains an h-segment γ_h and since $|l| \geq 2$, $G(\gamma_h)$ contains a v-segment (see Figure 7.3). Hence $H^i(\gamma_v)$ contains a v-segment for all $i \geq m$. Similarly backward iterations ensure that $H^{-j}(\gamma_v)$ contains a v-segment for all $j \geq n$. Therefore by Theorem 5.4.1 H has the Bernoulli property.   □

> **Key point:** This completes the proof of the Bernoulli property for toral linked twist maps.

7.7 Summary

The geometrical argument above gives sufficient conditions for a toral linked twist map to enjoy the Bernoulli property. The key point to notice is that different conditions are required for the co-rotating and counter-rotating cases. In particular, if a linked twist map is co-rotating, we only require the conditions from the previous chapter on the strength of the twists (that the twist functions should be monotonic). A wrapping number of 1 (i.e., single-twists) is sufficient. On the other hand, in the counter-rotating case, we require stronger conditions on the strength of the twist (given in Theorem 7.6.3), and also that we have a wrapping number of at least 2 (i.e., at least double-twists) to guarantee the Bernoulli property.

8

Linked twist maps on the plane

In this chapter we discuss the conversion of results for toral linked twist maps into results for planar linked twist maps. These systems seem more directly applicable to fluid mixing, but introduce new technical difficulties in the mathematics.

8.1 Introduction

In Chapter 2 we discussed the connection between linked twist maps and fluid flow, and observed that linked twist maps on the plane arise naturally in a number of existing experimental constructions, such as blinking flows and duct flows. However the extension of the results for toral linked twist maps to planar linked twist maps is not entirely straightforward. The situation for toral linked twist maps is relatively simple (at least in comparison to linked twist maps on other objects) because we can express twist maps in two independent directions in the same (Cartesian) coordinate system. The situation for planar annuli is more complicated.

As in Chapter 4, annuli in the plane and twist maps on such annuli are naturally described in polar coordinates. However, to create a linked twist map we require a pair of annuli with different centres. There is no simple coordinate system which then describes twist maps in both annuli. We therefore require additional transformations to move from one coordinate frame to another.

The following work in this chapter is mainly due to Wojtkowski (1980). This work predates Katok *et al.* (1986) by some six years, and so the author could not appeal to the Katok–Strelcyn version of Pesin theory for systems with singularities. Instead he constructs expanding and contracting segments of the tangent space explicitly, and proves their absolute continuity. Then he appeals to techniques in Anosov & Sinai (1967), Sinai (1970) and Weiss (1975) (which predate Pesin (1977)) to deduce the ergodic partition (and in fact the partition into K-components).

217

Thanks to the Katok–Strelcyn theorem, our work is easier, in that we do not have to concern ourselves with absolute continuity – this has been taken care of in Katok *et al.* (1986). As in Chapter 6 our strategy is to establish the existence of Lyapunov exponents, verify that any singularities allow the application of the Katok–Strelcyn theorem, and finally show that Lyapunov exponents are non-zero. To do so we use the construction of expanding and contracting sectors in the tangent space computed ingeniously in Wojtkowski (1980).

8.2 Planar linked twist maps

In this section we give the domain and the functions required to define a linked twist map on the plane, using the notation given in Wojtkowski (1980). Recall that in Chapter 4 we have already defined such a linked twist map using the notation of Devaney (1978).

8.2.1 The annuli

An annulus L in the plane $\mathbb{R}^2$ centered at the origin is most easily defined in the usual polar coordinates (r, θ) as

$$L = \{(r, \theta) : r_0 \leq r \leq r_1\}$$

where r_0, r_1 are positive constants. Recalling that Cartesian coordinates (x, y) can be expressed as $x = r \cos \theta$ and $y = r \sin \theta$ we see that we can translate L one unit to the left, that is, into an identical annulus centered at $(-1, 0)$ in Cartesian coordinates, by the transformation $M_1 : L \to \mathbb{R}^2$ given by

$$M_1(r, \theta) - (r \cos \theta - 1, r \sin \theta).$$

We give the annulus $M_1(L)$ the name R_1. Similarly a transformation $M_2 : L \to \mathbb{R}^2$ given by

$$M_2(r, \theta) = (1 - r \cos \theta, -r \sin \theta)$$

produces an annulus $R_2 = M_2(L)$ centred at $(1, 0)$.

The inverse transformations $M_1^{-1} : \mathbb{R}^2 \to L$ and $M_2^{-1} : \mathbb{R}^2 \to L$ are given by

$$M_1^{-1}(x, y) = \left(\sqrt{(x + 1)^2 + y^2}, \tan^{-1} \frac{y}{x + 1} \right)$$

and

$$M_2^{-1}(x, y) = \left(\sqrt{(1 - x)^2 + y^2}, \tan^{-1} \frac{y}{x - 1} \right).$$

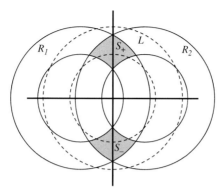

Figure 8.1 The annulus L is marked with dotted lines. The transformed annuli R_1 and R_2 are marked with solid lines and are centred on $(x, y) = (-1, 0)$ and $(1, 0)$ respectively. Because $r_0 > 1$ and $r_1 < 2 + r_0$ the intersection $R_1 \cap R_1$ contains two disconnected components, S_+ and S_-, shaded in grey.

Note that the transformations M_1 and M_2 are related by the transformation $C : \mathbb{R}^2 \to \mathbb{R}^2$ given by

$$C(x, y) = (-x, -y),$$

that is, the central symmetry transformation. Then $M_1 = CM_2$ and $M_2 = CM_1$. In other words, each annulus is the image of the other after reflection in both x and y axes. Throughout we use polar coordinates to describe points in the annulus L, and Cartesian coordinates to describe points in R_1 and R_2.

Choosing $r_0 > 1$ and $r_1 < 2 + r_0$ ensures that every circle contained in R_1 intersects transversely every circle contained in R_2, and that the intersection $R_1 \cap R_2$ contains two distinct components S_+ and S_-. We define the intersection of the annuli $S = R_1 \cap R_2 = S_+ \cup S_-$ and the union $R = R_1 \cup R_2$. See Figure 8.1.

8.2.2 The twist maps

We define the twist maps on our original annulus L using the usual polar coordinates. Define a pair of twist maps $F_1 : L \to L$ and $F_2 : L \to L$ by

$$F_1(r, \theta) = (r, \theta + f_1(r))$$
$$F_2(r, \theta) = (r, \theta + f_2(r))$$

where f_1 and f_2 are twist functions $f_{1,2} : [r_0, r_1] \to \mathbb{R}$ with $f_{1,2}(r_0) = f_{1,2}(r_1) = 0 \pmod{2\pi}$ to ensure continuity at the boundaries of the annuli. As usual we require the functions f_1, f_2 to be monotonic and C^2, and we assume without loss of generality the strengths of the twists to be positive:

$$\frac{df_1}{dr} \geq c_1 > 0 \quad \text{and} \quad \frac{df_2}{dr} \geq c_2 > 0.$$

We can fix wrapping numbers (defined as in Chapter 6) k and l for F_1 and F_2 respectively by setting $f_1(r_1) = 2\pi k$ and $f_2(r_1) = 2\pi l$. The twist functions f_1 and f_2 may, but need not, be identical.

To create a twist map on one of the linked annuli we first move points to the original annulus L, then perform the twist, and finally transform back to the offset annulus. A twist map defined for points in the annulus R_1 will therefore have the form $T_1 : R \to R$ where

$$T_1 = \begin{cases} M_1 F_1 M_1^{-1} & \text{if } (x, y) \in R_1 \\ Id & \text{if } (x, y) \in R_2 \backslash R_1, \end{cases}$$

so that as before points outside R_1 are left unchanged, and similarly for points in R_2 we have the map $T_2 : R \to R$ given by

$$T_2 = \begin{cases} M_2 F_2 M_2^{-1} & \text{if } (x, y) \in R_2 \\ Id & \text{if } (x, y) \in R_1 \backslash R_2. \end{cases}$$

Both twist maps F_1 and F_2 perform their twists in a counter-clockwise sense (since $f_1, f_2 > 0$). An inspection of the action of M_1 and M_2 will reveal that the twist maps T_1 and T_2 also act in a counter-clockwise sense. See Figure 8.2.

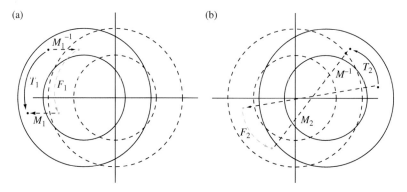

(a)   (b)

Figure 8.2 The action of twist maps T_1 and T_2. Figure (a) shows a point in R_1 mapped into L under M_1^{-1}, sheared under F_1 in L, and returned to R_1 under M_1. The composition $M_1 F_1 M_1^{-1}$ results in the transformation T_1, with the twist in the same sense as that of F_1. In Figure (b) the action of $T_2 = M_2 F_2 M_2^{-1}$ for a point in R_2 is shown. Note here that the effect of both M_2 and M_2^{-1} is to reflect in both the x and y axes as well of a translation of one unit to the right. The transformation T_2 is again in the same sense as that of F_2.

8.2.3 Linked twist maps

Now we can form a linked twist mapping by composing the twist maps, giving

$$T = T_2 \circ T_1.$$

Recalling the geometrical argument in Section 2.3.1 we note that this *co-rotating* linked twist map has a *counter-twisting* effect, since the dynamics in the inter-sections is locally equivalent to the dynamics in the counter-rotating toral linked twist map. A *counter-rotating/co-twisting* linked twist map can be formed by composing T_1 with the inverse of T_2:

$$\bar{T} = T_2^{-1} \circ T_1.$$

By analogy with the results of Chapters 6 and 7 we might expect that the co-rotating/counter-twisting system T to require stricter conditions on the twists to guarantee an ergodic partition than the counter-rotating/co-twisting version $\bar{T}$, and we shall see that this is indeed the case. For the rest of the chapter we work with compositions of many maps, and for ease of notation we will drop the composition symbol '$\circ$' and write, for example, $T = T_2 T_1$.

As in previous chapters we will work with a first return map T_S. Because the twist maps F_1 and F_2 are defined in polar coordinates, we will give the first return map in these coordinates. Thus for an initial condition $(r, \theta) \in M_1^{-1}(S)$ define $T_S : M_1^{-1}(S) \to M_1^{-1}(S)$ by

$$T_S(r, \theta) = M_1^{-1} T_2^m T_1^n M_1$$
$$= M_1^{-1} M_2 F_2^m M_2^{-1} M_1 F_1^n M_1^{-1} M_1$$
$$= M_1^{-1} M_2 F_2^m M_2^{-1} M_1 F_1^n$$

where n is such that $F_1^n(r, \theta) \in M_1^{-1}(S)$ and $F_1^i(r, \theta) \notin M_1^{-1}(S)$ for $1 \le i < n$, and similarly, m is such that $F_2^m(r, \theta) \in M_2^{-1}(S)$ and $F_2^i(r, \theta) \notin M_2^{-1}(S)$ for $1 \le i < m$. The first return map for $\bar{T}$, $\bar{T}_S : M_1^{-1}(S) \to M_1^{-1}(S)$ is defined exactly analogously:

$$\bar{T}_S(r, \theta) = M_1^{-1} T_2^{-m} T_1^n M_1$$
$$= M_1^{-1} M_2 F_2^{-m} M_2^{-1} M_1 F_1^n$$

where n is such that $F_1^n(r, \theta) \in M_1^{-1}(S)$ and $F_1^i(r, \theta) \notin M_1^{-1}(S)$ for $1 \le i < n$, and similarly, m is such that $F_2^{-m}(r, \theta) \in M_2^{-1}(S)$ and $F_2^{-i}(r, \theta) \notin M_2^{-1}(S)$ for $1 \le i < m$.

> **Key point:** Linked twist maps on the plane are defined using polar coordinates to express a twist on a central annulus, and a pair of coordinate-change transformations to move to the linked annuli in Cartesian coordinates.

8.3 The ergodic partition

To prove ergodicity results about T and $\bar{T}$ we follow the same scheme as in Chapter 6. We aim to show that all Lyapunov exponents, for almost every point in our domain, are non-zero, and then to appeal to the Katok–Strelcyn version of Pesin theory. We have defined the system in such a way as to make the two **(KS)** conditions trivially satisfied.

Lemma 8.3.1 *Planar linked twist maps T and $\bar{T}$ satisfy the conditions* **(KS1)** *and* **(KS2)**.

Proof $B(\mathrm{sing}T, \epsilon) = B(\partial R_1 \cup \partial R_2, \epsilon)$. An ϵ-neighbourhood around $\partial R_1 \cup \partial R_2$ consists of four annuli, two of inner radius $r_0 - \epsilon$ and outer radius $r_0 + \epsilon$, and two of inner radius $r_1 - \epsilon$ and outer radius $r_1 + \epsilon$. These annuli have areas $\pi(r_0 + \epsilon)^2 - \pi(r_0 - \epsilon)^2 = 4r_0\pi\epsilon$, and $\pi(r_1 + \epsilon)^2 - \pi(r_1 - \epsilon)^2 = 4r_1\pi\epsilon$ respectively. Thus $\mu(B(\mathrm{sing}T, \epsilon)) = 8\pi\epsilon(r_0 + r_1)$ and so Equation (5.24) is satisfied with $C_1 = 8\pi(r_0 + r_1)$ and $a = 1$. The same is true for $\bar{T}$.

The arguments in Assumption 6.4.1 apply here to show that T and $\bar{T}$ satisfy Equation (5.25) under very mild restrictions. $\square$

Moreover the Oseledec Multiplicative Ergodic Theorem again guarantees the existence of Lyapunov exponents for both T and T_S (and $\bar{T}$ and $\bar{T}_S$).

Lemma 8.3.2 *T, $\bar{T}$, T_S and $\bar{T}_S$ satisfy the Oseledec condition* **(OS)**.

Proof We have

$$\int_S \log^+ \|DT_S\| \mathrm{d}\mu \leq \int_R \log^+ \|DT\| \mathrm{d}\mu \tag{8.1}$$

$$= \int_{V=\mathrm{int}R\backslash\partial S} \log^+ \|DT\| \mathrm{d}\mu \tag{8.2}$$

$$< \infty \tag{8.3}$$

where the first inequality follows because $S \subset R$, the equality follows since $\mu(R\backslash V) = \mu(\mathrm{sing}T) = 0$, and the second inequality because T is C^2 on $V = \mathrm{int}R\backslash\partial S$. Again the proof is identical for $\bar{T}$ and $\bar{T}_S$. $\square$

All that remains is to show that the return maps T_S and $\bar{T}_S$ have non-zero Lyapunov exponents. We do so by using the complementary invariant cone technique discussed in Chapter 5 to derive Anosov conditions.

We use coordinates $(\beta_1, \beta_2) = (dr, d\theta)$ in the tangent space to L. Since $T_S = (M_1^{-1}M_2)F_2^m(M_2^{-1}M_1)F_1^n$ (and similarly for $\bar{T}_S$), when studying the action of DT_S on tangent vectors we need to consider the effect of DF_1^n, DF_2^m and DF_2^{-m} for $m, n \geq 1$, but also $D(M_2^{-1}M_1)$ and $D(M_1^{-1}M_2)$. The Jacobians DF_1^n, DF_2^m and DF_2^{-m} are straightforward to compute:

$$DF_1^n = \begin{pmatrix} 1 & 0 \\ nf_1' & 1 \end{pmatrix}, \quad DF_2^m = \begin{pmatrix} 1 & 0 \\ mf_2' & 1 \end{pmatrix}, \quad DF_2^{-m} = \begin{pmatrix} 1 & 0 \\ -mf_2' & 1 \end{pmatrix},$$

where the derivatives f_1' and f_2' are evaluated at the appropriate value of r. The Jacobians $D(M_2^{-1}M_1)$ and $D(M_1^{-1}M_2)$ require a little more effort. The composition $M_2^{-1}M_1$ is given by

$$M_2^{-1}M_1(r, \theta) = M_2^{-1}(r\cos\theta - 1, r\sin\theta)$$

$$= \left(\sqrt{4 - 4r\cos\theta + r^2}, \tan^{-1}\frac{r\sin\theta}{r\cos\theta - 2} \right).$$

Note that $M_1^{-1}M_2$ is identical, since $M_1^{-1}M_2 = M_1^{-1}CM_1 = M_2^{-1}M_1$.

Referring to Figure 8.3, let $(r, \theta) \in M_1^{-1}(S) \in L$. Then $M_1(r, \theta)$ has Cartesian coordinates $(x, y) = (r\cos\theta - 1, r\sin\theta) \in S$. Translating this point into R_2 we have, in polar coordinates, $M_2^{-1}M_1(r, \theta) = (\bar{r}, \bar{\theta})$ (where a simple calculation gives $\bar{r} = \sqrt{(1-x)^2 + y^2} = \sqrt{4 - 4r\cos\theta + r^2}$). Then by the chain rule,

$$DM_2^{-1}M_1|_{(r,\theta)}$$

$$= DM_2^{-1}|_{M_1(r,\theta)}DM_1|_{(r,\theta)}$$

$$= DM_2^{-1}|_{(x=r\cos\theta-1, y=r\sin\theta)}DM_1|_{(r,\theta)}$$

$$= \begin{pmatrix} \dfrac{x-1}{\sqrt{(1-x)^2 + (-y)^2}} & \dfrac{y}{\sqrt{(1-x)^2 + (-y)^2}} \\ \dfrac{-y}{(1-x)^2 + y^2} & \dfrac{x-1}{(1-x)^2 + y^2} \end{pmatrix} DM_1|_{(r,\theta)}$$

$$= \begin{pmatrix} \dfrac{r\cos-2}{\sqrt{(2-r\cos\theta)^2+(-r\sin\theta)^2}} & \dfrac{r\sin\theta}{\sqrt{(2-r\cos\theta)^2+(-r\sin\theta)^2}} \\ \dfrac{-r\sin\theta}{(2-r\cos\theta)^2+(-r\sin\theta)^2} & \dfrac{-(2-r\cos\theta)}{(2-r\cos\theta)^2+(-r\sin\theta)^2} \end{pmatrix}$$

$$\times \, DM_1|_{(r,\theta)}$$

$$= \begin{pmatrix} \dfrac{r\cos\theta-2}{\bar{r}} & \dfrac{r\sin\theta}{\bar{r}} \\ \dfrac{-r\sin\theta}{\bar{r}^2} & \dfrac{r\cos\theta-2}{\bar{r}^2} \end{pmatrix} \begin{pmatrix} \cos\theta & -r\sin\theta \\ \sin\theta & r\cos\theta \end{pmatrix}$$

$$= \begin{pmatrix} \dfrac{r-2\cos\theta}{\bar{r}} & \dfrac{2r\sin\theta}{\bar{r}} \\ \dfrac{-2\sin\theta}{\bar{r}^2} & \dfrac{r(r-2\cos\theta)}{\bar{r}^2} \end{pmatrix}.$$

Each element in this matrix can be expressed in terms of an angle $\alpha = \alpha(r,\theta)$, defined to be the angle between lines joining $(r,\theta) \in S$ with the centres of R_1 and R_2. In particular we have $\sin\alpha = 2\sin\theta/\bar{r}$ and $\cos\alpha = (r-2\cos\theta)/\bar{r}$. See Figure 8.3 and caption for details. We note as in Figure 8.3 that α, measured counter-clockwise, is positive in the upper half-plane, and negative in the lower half-plane. The range of α governs the size of the intersection domain S. Thus if $(r,\theta) \in M_1^{-1}(S_+)$ then

$$D(M_2^{-1}M_1)|_{(r,\theta)} = \begin{pmatrix} \cos\alpha(M_1(r,\theta)) & r\sin\alpha(M_1(r,\theta)) \\ -\sin\alpha(M_1(r,\theta))/\bar{r} & r\cos\alpha(M_1(r,\theta))/\bar{r} \end{pmatrix}$$

and if $(r,\theta) \in M_1^{-1}(S_-)$ then

$$D(M_2^{-1}M_1)|_{(r,\theta)} = \begin{pmatrix} \cos\alpha(M_1(r,\theta)) & -r\sin\alpha(M_1(r,\theta)) \\ \sin\alpha(M_1(r,\theta))/\bar{r} & r\cos\alpha(M_1(r,\theta))/\bar{r} \end{pmatrix}.$$

Similarly we have, for $(r,\theta) \in M_2^{-1}(S_+)$

$$D(M_1^{-1}M_2)|_{(r,\theta)} = \begin{pmatrix} \cos\alpha(M_2(r,\theta)) & r\sin\alpha(M_2(r,\theta)) \\ -\sin\alpha(M_2(r,\theta))/\bar{r} & r\cos\alpha(M_2(r,\theta))/\bar{r} \end{pmatrix}$$

and for $(r,\theta) \in M_2^{-1}(S_-)$

$$D(M_1^{-1}M_2)|_{(r,\theta)} = \begin{pmatrix} \cos\alpha(M_2(r,\theta)) & -r\sin\alpha(M_2(r,\theta)) \\ \sin\alpha(M_2(r,\theta))/\bar{r} & r\cos\alpha(M_2(r,\theta))/\bar{r} \end{pmatrix}.$$

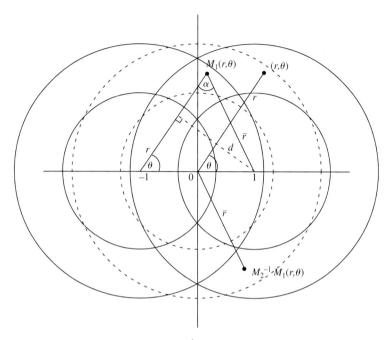

Figure 8.3 For any point $(r, \theta) \in M_1^{-1}(S_+)$ the image $M_1(r, \theta)$ lies in S_+. The angle $\alpha(M_1(r, \theta))$ is formed by subtending lines from $M_1(r, \theta)$ to $(-1, 0)$ and $(1, 0)$. The point $M_2^{-1} M_1(r, \theta)$ is defined to have polar radius $\bar{r}$, and so the lengths of the subtended lines are given as r and $\bar{r}$ respectively. To calculate α we add the perpendicular of length d, from $(1, 0)$ to the line subtended to $(-1, 0)$. Then $\sin \alpha = d/\bar{r}$. But we also have $\sin \theta = d/2$ and so $\sin \alpha = 2 \sin \theta / \bar{r}$. A similar calculation gives $\cos \alpha = (r - 2 \cos \theta)/\bar{r}$.

8.3.1 Counter-rotating planar linked twist maps

We turn first to the counter-rotating (co-twisting) linked twist map $\bar{T}$.
Let

$$\eta = \sup_{(r, \theta) \in M_1^{-1}(S)} \frac{|\cot \alpha(M_1(r, \theta))|}{r}$$

Theorem 8.3.1 (Wojtkowski (1980)) *The counter-rotating planar linked twist map* $T = T_2^{-1} T_1$ *is a union of (possibly countably many) ergodic components if*

$$\min(c_1, c_2) > 2\eta \tag{8.4}$$

We prove this theorem using the following lemmas. We aim to find a strictly invariant expanding cone field (recall Section 5.5), that is, a collection of cones in the tangent space at each point such that all vectors within each cone are expanded under $D\bar{T}_S$, and that each cone is mapped into the interior of the cone at the next iterate of the map. Taking the constituent mappings of $\bar{T}_S$ in turn we give cones U_1, $\tilde{U}_1$, U_2 and $\tilde{U}_2$ such that

$$DF_1^n(U_1) \subseteq \tilde{U}_1$$

$$D(M_2^{-1}M_1)(\tilde{U}_1) \subset U_2$$

$$DF_2^{-m}(U_2) \subseteq \tilde{U}_2$$

$$D(M_1^{-1}M_2)(\tilde{U}_2) \subset U_1$$

for $n, m > 0$. The fact that at least one of the inclusions above is strict is enough that $D\bar{T}_S(U_1) \subset U_1$.

Define first the cones

$$U_1 = \left\{ (\beta_1, \beta_2) : \frac{\beta_2}{\beta_1} \geq \frac{-c_1}{2} \right\},$$

$$\tilde{U}_1 = \left\{ (\beta_1, \beta_2) : \frac{\beta_2}{\beta_1} \geq \frac{c_1}{2} \right\}.$$

(Strictly speaking these cones are defined for each $(r, \theta) \in M_1^{-1}(S)$ so that $U_1 = U_1(r, \theta)$ etc. However since U_1 etc. will consist of the same tangent vectors at each point we write U_1 without confusion.)

Lemma 8.3.3

$$DF_1^n(U_1) \subseteq \tilde{U}_1$$

for each $n \geq 1$.

Proof Taking a point (r, θ) with tangent vector $v = (\beta_1, \beta_2) \in U_1$, we form $v' = (\beta_1', \beta_2') = DF_1^n(U_1)$ as follows:

$$\begin{pmatrix} \beta_1' \\ \beta_2' \end{pmatrix} = \begin{pmatrix} 1 & 0 \\ nf_1'(r) & 1 \end{pmatrix} \begin{pmatrix} \beta_1 \\ \beta_2 \end{pmatrix} = \begin{pmatrix} \beta_1 \\ nf_1'(r)\beta_1 + \beta_2 \end{pmatrix}$$

Then

$$\frac{\beta_2'}{\beta_1'} = \frac{nf_1'(r)\beta_1 + \beta_2}{\beta_1} \geq nc_1 + \frac{\beta_2}{\beta_1} \geq nc_1 + \frac{-c_1}{2} \geq \frac{c_1}{2}$$

for $n \geq 1$ and so $v' \in \tilde{U}_1$. Hence $DF_1^n(U_1) \subseteq \tilde{U}_1$. $\square$

Similarly we define the sectors

$$U_2 = \left\{ (\beta_1, \beta_2) : \frac{\beta_2}{\beta_1} \leq \frac{c_2}{2} \right\},$$

$$\tilde{U}_2 = \left\{ (\beta_1, \beta_2) : \frac{\beta_2}{\beta_1} \leq \frac{-c_2}{2} \right\}.$$

Lemma 8.3.4

$$DF_2^{-m}(U_2) \subseteq \tilde{U}_2$$

for each $m \geq 1$.

Proof Taking a point (r, θ) with tangent vector $v = (\beta_1, \beta_2) \in U_2$, we form $v' = (\beta_1', \beta_2') = DF_2^{-m}(U_2)$ as follows:

$$\begin{pmatrix} \beta_1' \\ \beta_2' \end{pmatrix} = \begin{pmatrix} 1 & 0 \\ -mf_2'(r) & 1 \end{pmatrix} \begin{pmatrix} \beta_1 \\ \beta_2 \end{pmatrix} = \begin{pmatrix} \beta_1 \\ -mf_2'(r)\beta_1 + \beta_2 \end{pmatrix}.$$

Then

$$\frac{\beta_2'}{\beta_1'} = \frac{-mf_2'(r)\beta_1 + \beta_2}{\beta_1} \leq -mc_2 + \frac{\beta_2}{\beta_1} \leq -mc_2 + \frac{c_2}{2} \leq \frac{-c_2}{2}$$

for $m \geq 1$ and so $v' \in \tilde{U}_2$. Hence $DF_2^{-m}(U_2) \subseteq \tilde{U}_2$. $\qquad\square$

We have not had to use the condition in Theorem 8.3.1 as these lemmas relate only to twist maps in the original annulus L. We will need the condition in the following lemmas.

Lemma 8.3.5 *Given condition (8.4) of Theorem 8.3.1 we have*

$$D(M_2^{-1}M_1)(\tilde{U}_1) \subset U_2.$$

Proof Let $M_2^{-1}M_1(r, \theta) = (\bar{r}, \bar{\theta})$. First we assume $(r, \theta) \in M_1^{-1}(S_+)$. We consider $D(M_2^{-1}M_1)(\tilde{U}_1)$ by taking $v = (\beta_1, \beta_2) \in \tilde{U}_1$, and compute $v' = (\beta_1', \beta_2') = D(M_2^{-1}M_1)v$. Then (to simplify the notation we write α instead of $\alpha(M_1(r, \theta))$),

$$\frac{\beta_2'}{\beta_1'} = \frac{-1/r\bar{r} + (\beta_2/\beta_1)(\cot\alpha/\bar{r})}{\cot\alpha/r + \beta_2/\beta_1}$$

$$< \frac{(\beta_2/\beta_1)(\cot\alpha/\bar{r})}{\cot\alpha/r + \beta_2/\beta_1} \qquad (8.5)$$

$$= \frac{\cot\alpha/\bar{r}}{(\beta_1/\beta_2)(\cot\alpha/r) + 1}$$

$$\leq \frac{\cot\alpha}{\bar{r}} \qquad (8.6)$$

$$< \frac{c_2}{2} \qquad (8.7)$$

In the above equations, inequality (8.5) holds provided the denominator, $\cot\alpha/r + \beta_2/\beta_1$ is positive (since r, $\bar{r} > 0$). But since $\cot\alpha/r + \beta_2/\beta_1 > \cot\alpha/r + c_1/2$, this holds by condition (8.4). In the case of inequality (8.6) there are two cases. If $\cot\alpha > 0$ we require $(\beta_1/\beta_2)(\cot\alpha/r) > 0$, but this is clearly true since $\beta_1/\beta_2 \geq c_1/2 > 0$. On the other hand if $\cot\alpha < 0$ then we require $0 < (\beta_1/\beta_2)(\cot\alpha/r) + 1 < 1$. But $(\beta_1/\beta_2)(\cot\alpha/r) > (2/c_1)(-c_1/2) = -1$ providing $c_1 > 2\eta$, which follows from condition (8.4). Finally inequality (8.7) follows directly from condition (8.4).

An almost identical procedure governs initial conditions in the other intersecting region. Assume $(r,\theta) \in M_1^{-1}(S_-)$. Then

$$
\begin{aligned}
\frac{\beta_2'}{\beta_1'} &= \frac{1/r\bar{r} + (\beta_2/\beta_1)(\cot\alpha/\bar{r})}{\cot\alpha/r - \beta_2/\beta_1} \\
&< \frac{(\beta_2/\beta_1)(\cot\alpha/\bar{r})}{\cot\alpha/r - \beta_2/\beta_1} \qquad\qquad\qquad (8.8) \\
&= \frac{\cot\alpha/\bar{r}}{(\beta_1/\beta_2)(\cot\alpha/r) - 1} \\
&\leq \frac{\cot\alpha}{\bar{r}} \qquad\qquad\qquad\qquad\quad (8.9) \\
&< \frac{c_2}{2} \qquad\qquad\qquad\qquad\qquad (8.10)
\end{aligned}
$$

Similarly to the above, inequality (8.8) holds since condition (8.4) implies $\cot\alpha/r - \beta_2/\beta_1 < 0$. To verify inequality (8.9) we consider two cases. If $\cot\alpha > 0$ we require $(\beta_1/\beta_2)(\cot\alpha/r) - 1 < 0$, but this holds because $(\beta_1/\beta_2)(\cot\alpha/r) < (2/c_1)(c_1/2) = 1$, by condition (8.4). If $\cot\alpha < 0$ we require $(\beta_1/\beta_2)(\cot\alpha/r) - 1 < -1$. But this is clear since $\beta_1/\beta_2 > 0$. Finally inequality (8.10) is given by condition (8.4). $\qquad\square$

An almost identical proof applies to the final inclusion lemma:

Lemma 8.3.6 *Given condition (8.4) of Theorem 8.3.1 we have*

$$D(M_1^{-1}M_2)(\tilde{U}_2) \subset U_1$$

Combining the previous four lemmas we have $D\bar{T}_S(U_1) \subset U_1$. Thus the cone field $U_+ = \cup_{(r,\theta)\in M_1^{-1}(S)} U_1$ is strictly invariant, so that $D\bar{T}_S(U_+) \subset U_+$. Now we must show that $D\bar{T}_S$ expands vectors in U_1. Since we are working in polar coordinates we use the usual Riemannian metric $ds^2 = dr^2 + r^2d\theta^2$, so that the norm of a vector (β_1, β_2) in the tangent space at a point (r,θ) is $\sqrt{\beta_1^2 + r^2\beta_2^2}$.

The mapping $M_2^{-1}M_1$ results in no expansion or contraction of tangent vectors, as shown in the following lemma.

Lemma 8.3.7 *Given a tangent vector $v = (\beta_1, \beta_2) \in T_p L$ we have*

$$\|D(M_2^{-1}M_1)v\| = \|v\|.$$

Proof Let $(\bar{r}, \bar{\theta}) = M_2^{-1} M_1(r, \theta)$. Then, writing α for $\alpha(M_1(r, \theta))$, we have

$$D(M_2^{-1}M_1)v = \begin{pmatrix} \cos \alpha & \pm r \sin \alpha \\ \mp \sin \alpha / \bar{r} & r \cos \alpha / \bar{r} \end{pmatrix} \begin{pmatrix} \beta_1 \\ \beta_2 \end{pmatrix}$$

$$= \begin{pmatrix} \beta_1 \cos \pm \beta_2 r \sin \alpha \\ \mp \beta_1 \sin \alpha / \bar{r} + \beta_2 r \cos \alpha / \bar{r} \end{pmatrix}$$

and so

$$\|D(M_2^{-1}M_1)v\| = \sqrt{(\beta_1 \cos \alpha \pm \beta_2 r \sin \alpha)^2 + \bar{r}^2(\mp \beta_1 \sin \alpha / \bar{r} + \beta_2 r \cos \alpha / \bar{r})^2}$$

$$= \sqrt{\beta_1^2 + r^2 \beta_2^2}$$

$$= \|v\|.$$

$\square$

An identical result holds for $D(M_1^{-1}M_2)$. These are to be expected as M_1 and M_2 are simply translations, and as such have no expanding or contracting properties. The expansion and contraction comes, as before, from the twist maps.

Lemma 8.3.8 F_1 *increases the norm of all vectors from U_1 except for vectors of the form $(0, \beta_2)$, that is,*

$$\|DF_1 v\| > \|v\|,$$

for $v \in U_1$ except $v = (0, \beta_2)$.

Proof Consider a point (r, θ) and a tangent vector $v = (\beta_1, \beta_2) \in U_1$ so that $\|v\| = \|(\beta_1, \beta_2)\| = \sqrt{\beta_1^2 + r^2 \beta_2^2}$. Let $F_1(r, \theta) = (r, \bar{\theta})$ (recall that F_1 leaves r unchanged). Then

$$\|DF_1 v\| = \sqrt{\beta_1^2 + r^2(f'\beta_1 + \beta_2)^2}$$

$$= \sqrt{\beta_1^2 + r^2 \beta_2^2 + r^2 f'(f'\beta_1^2 + 2\beta_1 \beta_2)}.$$

But $f'\beta_1^2 + 2\beta_1\beta_2 > c_1\beta_1^2 + 2\beta_1\beta_2 \geq 0$ (provided $\beta_1 \neq 0$) since $(\beta_1, \beta_2) \in U_1$. Hence $\|DF_1(\beta_1, \beta_2)\| > \|(\beta_1, \beta_2)\|$ unless $(\beta_1, \beta_2) = (0, \beta_2)$. $\square$

Lemma 8.3.9 F_2^{-1} *increases the norm of all vectors from* U_2 *except for vectors of the form* $(0, \beta_2)$, *that is,*

$$\|DF_2^{-1}v\| > \|v\|,$$

for $v \in U_2$ *except* $v = (0, \beta_2)$.

Proof Similarly to the previous proof consider a point (r, θ) with a tangent vector $v = (\beta_1, \beta_2) \in U_2$ so that $\|v\| = \sqrt{\beta_1^2 + r^2 \beta_2^2}$. Let $F_2^{-1}(r, \theta) = (r, \bar{\theta})$. Then

$$\|DF_2^{-1}v\| = \sqrt{\beta_1^2 + r^2(-f'\beta_1 + \beta_2)^2}$$

$$= \sqrt{\beta_1^2 + r^2 \beta_2^2 + r^2 f'(f'\beta_1^2 - 2\beta_1\beta_2)}.$$

But $f'\beta_1^2 - 2\beta_1\beta_2 > c_2\beta_1^2 - 2\beta_1\beta_2 \geq 0$ (provided $\beta_1 \neq 0$) since $(\beta_1, \beta_2) \in U_2$. Hence $\|DF_2^{-1}(\beta_1, \beta_2)\| > \|(\beta_1, \beta_2)\|$ unless $(\beta_1, \beta_2) = (0, \beta_2)$. $\qquad\square$

To take account of vectors of the form $(0, \beta_2)$ we give the following lemma:

Lemma 8.3.10 *Vectors of the form* $(0, \beta_2)$ *are mapped into vectors of the form* $(\beta_1 \neq 0, \beta_2)$ *under* $D(M_2^{-1}M_1)$ *and* $D(M_1^{-1}M_2)$.

Proof Let $v = (0, \beta_2)$. Then

$$D(M_2^{-1}M_1)v = \begin{pmatrix} \cos\alpha & \pm r\sin\alpha \\ \mp\sin\alpha/\bar{r} & r\cos\alpha/\bar{r} \end{pmatrix}\begin{pmatrix} 0 \\ \beta_2 \end{pmatrix}$$

$$= \begin{pmatrix} \pm r\beta_2\sin\alpha \\ (r\beta_2\cos\alpha)/\bar{r} \end{pmatrix}.$$

Since $r > 0$ and $\beta_2 \neq 0$, the first component of this is equal to zero if and only if $\sin\alpha = 0$. But for this we would require $\alpha = 0$ or $\pm\pi$, which cannot happen since $r_0 > 1$ and $r_2 < 2 + r_0$. $\qquad\square$

Now we can combine these to deduce expansion under the linked twist map $\bar{T}$.

Lemma 8.3.11 *For vectors* $v \in U_+$

$$\|D\bar{T}_S v\| \geq \lambda\|v\| \qquad\qquad (8.11)$$

where $\lambda > 1$.

Proof Referring back to Proposition 5.5.1, we have given a cone field U_+ which is mapped strictly into itself under $\bar{T}_S$. Moreover $\bar{T}_S$ expands all vectors $v \in U_+$ (F_1 expands all vectors except those of the form $(0, \beta_2)$. Vectors of that form are mapped by DF_1 into vectors of the form $(\beta_0 \neq 0, \beta_2)$. Then F_2 expands all vectors except those of the form $(0, \beta_2)$, and so in particular expands those vectors not expanded by F_1). $\qquad \Box$

In a precisely analogous way we can construct a complementary cone field U_- such that $D\bar{T}_S^{-1}(U_-) \subset U_-$ and such that $D\bar{T}_S^{-1}$ expands all vectors in U_-.

Lemma 8.3.12 *For vectors* $v \in U_-$

$$\|D\bar{T}_S^{-1}v\| \geq \mu\|v\| \qquad (8.12)$$

where $\mu > 1$.

This gives the now familiar result that the first return map $\bar{T}_S$ to the intersection region S is uniformly hyperbolic. The original map $\bar{T}$ is nonuniformly hyperbolic as different points take different numbers of iterations to return to S. As in Chapter 6 our desired result follows directly.

Proof of Theorem 8.3.1 Lemma 8.3.1 shows that singularities in the system are such that we can appeal to the Katok–Strelcyn theory. Lemma 8.3.2 shows that Lyapunov exponents exist for both T and the first return map T_S. The Anosov conditions given by Equations (8.11) and (8.12) in Lemmas 8.3.11 and 8.3.12 guarantee that Lyapunov exponents for $\bar{T}_S$ are non-zero. Our usual argument applies to show that $\bar{T}$ itself has non-zero Lyapunov exponents. Then the result follows by the Katok–Strelcyn version of Pesin theory (Theorem 5.4.1). $\qquad \Box$

Example 8.3.1 *Figure 8.4 shows an example of a planar counter-rotating linked twist map $\bar{T}$ which satisfies the conditions of Theorem 8.3.1. Here the inner and outer radii r_0 and r_1 equal 2 and 3 respectively. We have single twists $(k = l = 1)$ and the twist functions f_1 and f_2 are linear. Two initial blobs of 10,000 points are shown in the first picture. The following pictures are images of these initial blobs under 1, 2, 3, 5 and 10 iterates of $\bar{T}$. This example satisfies the conditions of Theorem 8.3.1 since $c_1 = c_2 = 2\pi$, while $\eta \approx 0.383$.*

Example 8.3.2 *Figure 8.5 shows an example of a planar counter-rotating linked twist map $\bar{T}$ which does not satisfy the conditions of Theorem 8.3.1 but still appears (numerically) to mix well. Here the inner and outer radii r_0 and r_1 equal 1.1 and 3 respectively, and the intersecting region S is much larger than in the previous example. We have single twists $(k = l = 1)$ and the twist functions f_1 and f_2 are linear. Two initial blobs of 10,000 points are shown in the first picture. The following pictures are images of these initial blobs under 1, 2, 3, 5 and 10 iterates of $\bar{T}$. However in this example we have $c_1 = c_2 = 2\pi/1.99 \approx 3.16$, while $2\eta \geq 6.84$. The map still appears to mix well despite violating the*

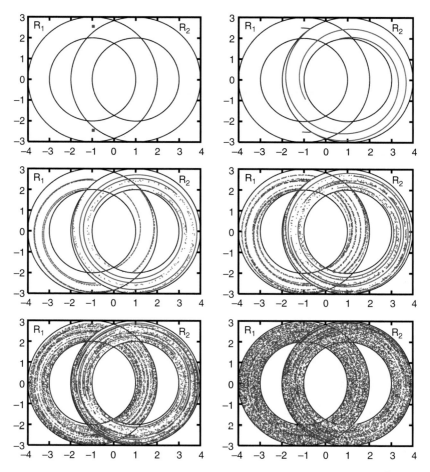

Figure 8.4 Successive iterates of the counter-rotating planar linked twist map $\bar{T}$, with $r_0 = 2$, $r_1 = 3$, and $k = l = 1$. The twist functions f_1 and f_2 are linear. Initially (in the top left picture) we have blobs consisting of 10,000 points. The next five pictures are the images of these blobs under the iterates 1, 2, 3, 5, 10 of the map.

conditions of the theorem. Moreover, the domain of mixed fluid is much larger. This suggests, numerically, that the conditions in the theorem are not sharp.

Key point: The ergodic partition for planar twist maps can be deduced in much the same way as on the torus. Expanding and contracting cone fields can be constructed which guarantee non-zero Lyapunov exponents.

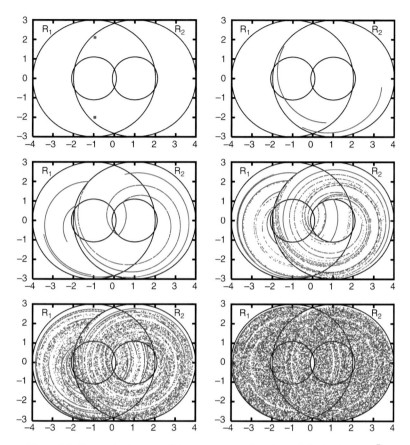

Figure 8.5 Successive iterates of the counter-rotating planar linked twist map $\bar{T}$, with $r_0 = 1.1$, $r_1 = 3$, and $k = l = 1$. The twist functions f_1 and f_2 are linear. Initially (in the top left picture) we have blobs consisting of 10,000 points. The next five pictures are the images of these blobs under the iterates 1, 2, 3, 5, 10 of the map.

8.3.2 Co-rotating planar linked twist maps

Theorem 8.3.2 *The co-rotating planar linked twist map $T = T_2 T_1$ is a union of (possibly countably many) ergodic components if*

$$c_1 c_2 - 2(c_1 + c_2)\eta > 4/r_0^2 > 0. \tag{8.13}$$

This is a more restrictive condition than that of Theorem 8.3.1. To see this observe that $c_1 c_2 - 2(c_1 + c_2)\eta = c_1(c_2 - 2\eta) - 2c_2\eta > 0 \implies c_2 - 2\eta > 0$, and similarly $c_1 - 2\eta > 0$. We proceed in the same manner as in the previous section, defining the cones

$$U_1 = \left\{ (\beta_1, \beta_2) : \frac{\beta_2}{\beta_1} \geq \frac{-c_1}{2} \right\},$$

$$\tilde{U}_1 = \left\{ (\beta_1, \beta_2) : \frac{\beta_2}{\beta_1} \geq \frac{c_1}{2} \right\},$$

$$U_2 = \left\{ (\beta_1, \beta_2) : \frac{\beta_2}{\beta_1} \geq \frac{-c_2}{2} \right\},$$

$$\tilde{U}_2 = \left\{ (\beta_1, \beta_2) : \frac{\beta_2}{\beta_1} \geq \frac{c_2}{2} \right\}.$$

Note that U_1 and $\tilde{U}_1$ are identical to previously, whereas U_2 and $\tilde{U}_2$ have changed. Thus Lemma 8.3.3 still applies, but we need a new lemma to show inclusion under DF_2^m.

Lemma 8.3.13

$$DF_2^m(U_2) \subseteq \tilde{U}_2$$

for each $m \geq 1$.

Proof Taking a point (r, θ) with tangent vector $v = (\beta_1, \beta_2) \in U_2$, we form $v' = (\beta_1', \beta_2') = DF_2^m(U_2)$ as follows:

$$\begin{pmatrix} \beta_1' \\ \beta_2' \end{pmatrix} = \begin{pmatrix} 1 & 0 \\ mf_2'(r) & 1 \end{pmatrix} \begin{pmatrix} \beta_1 \\ \beta_2 \end{pmatrix} = \begin{pmatrix} \beta_1 \\ mf_2'(r)\beta_1 + \beta_2 \end{pmatrix}.$$

Then

$$\frac{\beta_2'}{\beta_1'} = \frac{mf_2'(r)\beta_1 + \beta_2}{\beta_1} \geq mc_2 + \frac{\beta_2}{\beta_1} \geq mc_2 - \frac{c_2}{2} \geq \frac{c_2}{2}$$

for $m \geq 1$ and so $v' \in \tilde{U}_2$. Hence $DF_2^m(U_2) \subseteq \tilde{U}_2$. $\square$

The new condition in Theorem 8.3.2 (Equation (8.13)) is required for the inclusion lemmas for $D(M_2^{-1}M_1)$ and $D(M_1^{-1}M_2)$.

Lemma 8.3.14 *Given condition (8.13) of Theorem 8.3.2 we have*

$$D(M_2^{-1}M_1)(\tilde{U}_1) \subset U_2.$$

Proof Let $M_2^{-1}M_1(r, \theta) = (\bar{r}, \bar{\theta})$. First assume $(r, \theta) \in S_+$. Let $v = (\beta_1, \beta_2) \in \tilde{U}_1$ and let $v' = (\beta_1', \beta_2') = D(M_2^{-1}M_1)v$. Then as in Lemma 8.3.5 we have

$$\frac{\beta_2'}{\beta_1'} = \frac{-1/r\bar{r} + (\beta_2/\beta_1)(\cot\alpha/\bar{r})}{\cot\alpha/r + \beta_2/\beta_1}$$

But Equation (8.13) gives

$$c_1 c_2 - 2(c_1 + c_2)\eta > 4/r_0^2$$

$$\implies -1/r_0^2 + c_1 c_2/4 > \eta(c_2/2 + c_1/2)$$

$$\implies \frac{-1}{r\bar{r}} + \frac{\beta_2}{\beta_1}\frac{c_2}{2} > \eta\left(\frac{c_2}{2} + \frac{\beta_2}{\beta_1}\right) \tag{8.14}$$

$$\implies \frac{-1}{r\bar{r}} + \frac{\beta_2}{\beta_1}\frac{c_2}{2} > -\frac{\cot\alpha}{\bar{r}}\left(\frac{c_2}{2} + \frac{\beta_2}{\beta_1}\right) \tag{8.15}$$

$$\implies \frac{-1}{r\bar{r}} + \frac{\beta_2}{\beta_1}\frac{\cot\alpha}{\bar{r}} > -\frac{c_2}{2}\left(\frac{\cot\alpha}{r} - \frac{\beta_2}{\beta_1}\right)$$

$$\implies \frac{-1/r\bar{r} + (\beta_2/\beta_1)(\cot\alpha/\bar{r})}{\cot\alpha/r - \beta_2/\beta_1} > -\frac{c_2}{2} \tag{8.16}$$

$$\implies \frac{\beta_2'}{\beta_1'} > -\frac{c_2}{2}$$

where inequality (8.14) follows since $\beta_2/\beta_1 \geq c_1/2$ and $r, \bar{r} \geq r_0$, inequality (8.15) follows by the definition of η, and inequality (8.16) follows since $\cot\alpha/r - \beta_2/\beta_1 > c_1/2 - c_1/2 = 0$. A similar argument applies to $(r, \theta) \in S_-$. □

The final inclusion lemma follows almost identically.

Lemma 8.3.15 *Given condition (8.13) of Theorem 8.3.2 we have*

$$D(M_1^{-1}M_2)(\tilde{U}_2) \subset U_1.$$

Having shown the required inclusion properties we must prove the expansion of vectors for U_2 (the expansion of vectors in U_1 is shown by Lemma 8.3.8).

Lemma 8.3.16 *F_2 increases the norm of all vectors from U_2 except for vectors of the form $(0, \beta_2)$.*

Proof Consider a point (r, θ) with a tangent vector $v = (\beta_1, \beta_2) \in U_2$ so that $\|v\| = \sqrt{\beta_1^2 + r^2\beta_2^2}$. Let $F_2^{-1}(r, \theta) = (r, \bar{\theta})$. Then

$$\|DF_2 v\| = \sqrt{\beta_1^2 + r^2(f'\beta_1 + \beta_2)^2}$$

$$= \sqrt{\beta_1^2 + r^2\beta_2^2 + r^2 f'(f'\beta_1^2 + 2\beta_1\beta_2)}.$$

But $f'\beta_1^2 + 2\beta_1\beta_2 > c_2\beta_1^2 + 2\beta_1\beta_2 \geq 0$ (provided $\beta_1 \neq 0$) since $(\beta_1, \beta_2) \in U_2$. Hence $\|DF_2(\beta_1, \beta_2)\| > \|(\beta_1, \beta_2)\|$ unless $(\beta_1, \beta_2) = (0, \beta_2)$.   □

Since Lemma 8.3.10 holds for the counter-rotating case, the counterpart to Lemma 8.3.11 can be given:

Lemma 8.3.17 *For vectors $v \in U_+$*

$$\|DT_S v\| \geq \lambda \|v\| \tag{8.17}$$

where $\lambda > 1$.

Proof See the proof of lemma 8.3.11.   □

Again a complementary sector bundle U_- such that $DT_S^{-1}(U_-) \subset U_-$ and such that DT_S^{-1} expands all vectors in U_- produces

Lemma 8.3.18 *For vectors $v \in U_-$*

$$\|DT_S v\| \geq \mu \|v\| \tag{8.18}$$

where $\mu > 1$.

Proof of Theorem 8.3.2 Lemma 8.3.1 shows that singularities in the system are such that we can appeal to the Katok–Strelcyn theory. Lemma 8.3.2 shows that Lyapunov exponents exist for both T and the first return map T_S. The Anosov conditions given by Equations (8.17) and (8.18) in Lemmas 8.3.17 and 8.3.18 guarantee that Lyapunov exponents for T_S are non-zero. Our usual argument applies to show that T itself has non-zero Lyapunov exponents. Then the result follows by the Katok–Strelcyn version of Pesin theory (Theorem 5.4.1).   □

Example 8.3.3 *Figure 8.6 shows an example of a planar co-rotating linked twist map T which satisfies the conditions of Theorem 8.3.2. Here the inner and outer radii r_0 and r_1 equal 2 and 3 respectively. We have single twists ($k = l = 1$) and the twist functions f_1 and f_2 are linear. Two initial blobs of 10,000 points are shown in the first picture. The following pictures are images of these initial blobs under 1, 2, 3, 5 and 10 iterates of T. This example satisfies the conditions of Theorem 8.3.2 since $c_1 = c_2 = 2\pi$, while $\eta \approx 0.383$.*

Example 8.3.4 *Figure 8.7 shows an example of a planar co-rotating linked twist map T which does not satisfy the conditions of Theorem 8.3.2 but still appears (numerically) to mix well. Here the inner and outer radii r_0 and r_1 equal 1.1 and 3 respectively. We have single twists ($k = l = 1$) and the twist functions f_1 and f_2 are linear. Two initial blobs of 10,000 points are shown in*

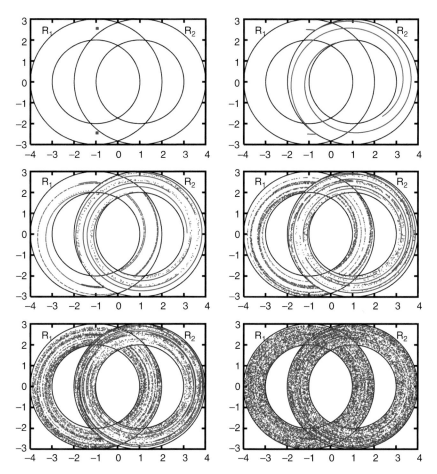

Figure 8.6 Successive iterates of the co-rotating planar linked twist map T, with $r_0 = 2$, $r_1 = 3$, and $k = l = 1$. The twist functions f_1 and f_2 are linear. Initially (in the top left picture) we have blobs consisting of 10,000 points. The next five pictures are the images of these blobs under the iterates 1, 2, 3, 5, 10 of the map.

the first picture. The following pictures are images of these initial blobs under 1, 2, 3, 5 and 10 iterates of T.

Key point: Just as for toral linked twist maps, stronger conditions are needed to guarantee the ergodic partition for the counter-twisting case than for the co-twisting case. On the plane the necessary conditions relate strength of twist to the cotangent of the angle α.

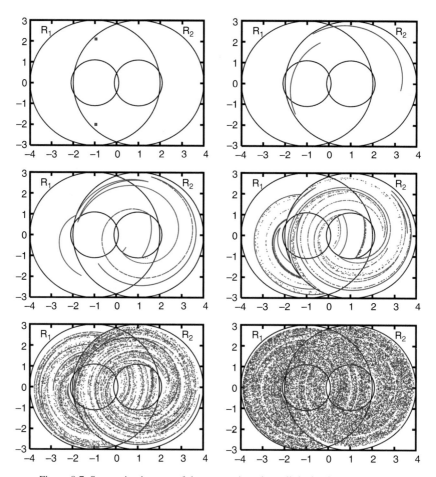

Figure 8.7 Successive iterates of the co-rotating planar linked twist map T, with $r_0 = 1.1$, $r_1 = 3$, and $k = l = 1$. The twist functions f_1 and f_2 are linear. Initially (in the top left picture) we have blobs consisting of 10,000 points. The next five pictures are the images of these blobs under the iterates 1, 2, 3, 5, 10 of the map.

8.4 Ergodicity and the Bernoulli property for planar linked twist maps

The geometrical argument of Wojtkowski (1980) and Przytycki (1983) discussed in Chapter 7 cannot be applied in the planar linked twist map scenario. Recall that in the toral case, h-segments and v-segments can be shown to be inevitable by considering the fact that in the co-rotating case, acute angles between segments of unstable manifold are not introduced, and in the counter-rotating case, taking sufficiently strong twists allows the intricate argument of Przytycki (1983). In the planar case it is difficult to deduce the eventual existence of h-segments and v-segments (defining such line segments in the appropriate way on S_+ and S_-) because, as pointed out by Wojtkowski (1980), the Jacobians DT and $D\bar{T}$ can change the orientation of line segments. Thus acute angles between line segments may be introduced, and a more sophisticated argument, possibly along the lines of Przytycki (1983), but necessarily more geometrically complicated, would be required. However it is perhaps reasonable, based on numerical experiments and an understanding of the behaviour of segments of stable and unstable manifolds to make the following conjecture.

Conjecture 8.4.1 *It is conjectured by Wojtkowski (1980), and this is supported numerically, that under the conditions of Theorem 8.3.1 (resp. Theorem 8.3.2), the linked twist map $\bar{T}$ (resp. T) is ergodic.*

If this is the case, and there are indeed integers m and n such that $T^m(\gamma^u(z)) \cap T^{-n}(\gamma^s(z)) \neq \emptyset$ (and similarly for $\bar{T}$) for local unstable and stable manifolds at almost every $z \in R$, then the corresponding result for the Bernoulli property follows naturally for the case with at least double twists.

Corollary 8.4.1 *If a planar linked twist map is ergodic, and has $k \geq 2$ and $l \geq 2$, then it possesses the Bernoulli property.*

Proof Just as in Theorem 7.6.3, as soon as we have an h-segment or a v-segment, the (at least) double twists of a linked twist map with $k \geq 2$ and $l \geq 2$ guarantees that the h- or v-segment is wrapped around the entire annulus (at least) twice, guaranteeing an h- or v-segment as appropriate at every iteration.   □

8.5 Summary

The above arguments give conditions for the co-rotating and counter-rotating planar linked twist maps to have an ergodic partition. Again there is a difference between the strictness of the conditions in the two cases. Numerical experiments appear to suggest that stronger results may hold. For example, just as in the toral case, there seems to be only one component in the ergodic decomposition (that is, the system is ergodic). However, the global argument applied to the toral case in Chapter 6 cannot be applied in the more geometrically complicated planar case. Moreover, the conditions for the theorems for both co- and counter-rotating planar linked twist maps to hold do not appear to be optimal. Examples have been given which violate the theorems for both cases, which still appear to mix well.

9

Further directions and open problems

We conclude by identifying some directions in which this work could and should be extended. In particular we discuss issues of optimization of size of mixing regions, lack of transversality in streamline crossing, and breakdown of monotonicity for twist functions.

9.1 Introduction

In this final chapter we discuss some of the open questions connected with the linked twist map approach to mixing. It is apparent that in terms of designing, creating and optimizing mixers this approach is still very much in its infancy. Translating the ideas of previous chapters into 'design' principles is itself a size-able task, and may result in the theory being extended in a variety of interesting directions. We mention only a few here.

In Section 9.2 we consider the question of how to optimize the size of domain on which the Bernoulli property is present. We have seen in previous chapters that, providing certain conditions on the properties of the twists are satisfied, the Bernoulli property is enjoyed by almost all points in the union of the annuli. However, we have also seen that the required conditions on the twists are more likely to be satisfied for smaller size domains. For example, recall that a counter-rotating toral linked twist map with linear single-twists is periodic with period 6 when the annuli are equal in size to the torus (see Example 6.4.2), while a linear counter-rotating TLTM may be Bernoulli for a choice of smaller annuli. We formalize this dichotomy quantitatively by computing, for both toral and planar linked twist maps with linear twists, the relationship between the area of intersection and union of the annuli, and the width of annuli required to guarantee that the system has an ergodic partition, is ergodic, and possesses the Bernoulli property. These calculations are carried out for the case of linear twists for simplicity. The case of nonlinear twists can be treated similarly. However the appropriate conditions relating the properties of the twist and geometry will require numerical computation.

The existing theorems discussed previously have some clear limitations, in that they require some conditions on the form of the twist functions and, in the planar case, on the size and positioning of the annuli. Moreover these conditions may be hard to fulfil in physical applications. For example, in the case of planar linked twist maps, we had the requirement that the inner and outer radii of the annuli were such that the intersection was composed of two disconnected regions. In many applications it may be that this restriction is not naturally met. In Section 9.3 we discuss the fact that the main issue in this instance is that of a lack of *transversality* (i.e., breakdown of 'crossing of streamlines'). We detail where the results of Chapter 8 break down, and give some simple numerical examples which suggest some obvious conjectures.

Another condition we have insisted upon throughout book is that the twist functions should be monotonic. Again, it may be that in applications this is not the case, for example because of boundary conditions or the rheological behaviour of the fluid in question. In Section 9.4 we discuss this issue, again explaining where the existing proofs fail. We give additional numerical experiments illustrating the behaviour that can arise in systems with nonmonotonic twist functions, which again suggest some natural conjectures. Both the issues of transversality and monotonicity are fundamental to hyperbolicity, and in turn, to mixing properties.

Finally in this chapter we close by mentioning several directions in which this work could be extended, and some areas of current research which may prove to have fruitful links to the theory of linked twist maps.

9.2 Optimizing mixing regions for linked twist maps

The results of the last few chapters have established conditions for linked twist maps to have an ergodic partition, in the case of toral and planar linked twist maps, and to enjoy the Bernoulli property, in the toral case. An immediate observation is that the properties of ergodicity, mixing and Bernoulli are defined in infinite time. In this sense they are necessary, but not sufficient, conditions for 'good mixing' in 'finite time'. Practically, we are more likely to be interested in the speed of mixing than in infinite time limits. As discussed in Chapter 3, and touched on again, briefly, at the end of this chapter, this question is usually discussed in the realm of *decay of correlations*, which are beyond the scope of this book. However, there may be other constraints of which to be aware. In applications we are likely to have a set of physical criteria, governed by the particular mechanism which creates the linked twist map structure. For example, it may be of paramount importance to achieve mixing on the largest possible domain, and so be desirable to maximize the area of the union of the annuli. There may be space constraints in the geometry of a particular mixer which necessitates making the area of intersection of the annuli as small as possible. It may be that the most important factor governing the efficiency of the mixer is the strength of the twist, or perhaps the wrapping number (defined in Chapter 6).

A complete survey of comparisons between different device criteria would likely fill another volume. As an illustration, in this section we concentrate on comparing the sizes of the intersection and union of the annuli, with the strength of the twists. First for the toral case, and then the planar case, we consider the twist functions as linear twists, in which case the strength of the twists are directly related to the width of the annuli, and then investigate the mixing properties of linked twist maps of different size. It can be seen by referring back to the theorems that, roughly speaking, the play-off is between increasing the size of the domain, which decreases the strength, but makes mixing less likely, and choosing the strength large enough to guarantee mixing, but which reduces the size of domain.

9.2.1 Toral linked twist maps

Co-rotating toral linked twist maps

Recalling the results of Chapters 6 and 7, we observe that the co-rotating toral linked twist map (TLTM) requires few constraints to guarantee the Bernoulli property on the whole region R. More precisely, a co-rotating TLTM composed of (k, α) and (l, β) twists is Bernoulli if $\alpha\beta > 0$ (see Theorems 6.4.1 and 7.6.2).

Counter-rotating toral linked twist maps

Counter-rotating toral linked twist maps are more interesting from the point of view of optimizing the size of the mixing region, as they are required to satisfy the inequality $|\alpha\beta| > 4$ (see Theorem 6.4.2) to guarantee an ergodic partition, and the stricter inequality $|\alpha\beta| > C \approx 17.24445$ (see Theorem 7.6.3), plus the condition of double-twists, that $|k|, |l| \geq 2$ to guarantee that the TLTM have the Bernoulli property.

For this simple illustration we consider a special case of the general counter-rotating toral linked twist map discussed in Chapter 6, which we may call a 'symmetric linear TLTM'. We assume we have a horizontal annulus P with $y_1 = 1$, and set $y_0 = r$, a 'radius parameter' which we shall vary to produce annuli of different widths. We have a corresponding vertical annulus Q with $x_1 = 1$ and $x_0 = r$, so that the annuli P and Q are equal in size. We choose identical (but with opposite sign) linear twists for *both* annuli, so that the twist on P is a (k, α) twist (with $k, \alpha > 0$), and the twist on Q is a $(-k, -\alpha)$ twist, recalling that in this case (see Section 6.2.8) the values of the derivatives (the strengths of the twists) are $|\alpha| = |k|/(1-r)$, where k is the wrapping number of the twists. Thus it is clear that the strength of a linear twist is entirely governed by the size of the annuli, for a given wrapping number.

In this simple symmetric case the conditions for the counter-rotating TLTM translate to

$$\alpha^2 = \frac{k^2}{(1-r)^2} \geq 4, \tag{9.1}$$

for the ergodic partition and

$$\alpha^2 = \frac{k^2}{(1-r)^2} \geq C \approx 17.24445, \tag{9.2}$$

together with $k \geq 2$ for the Bernoulli property.

The quantities to be optimized (the sizes of the union and intersection of the annuli) are easily computed. The area of $S = P \cap Q$ is

$$A_\cap(r) = (1-r)^2$$

while the area of $R = P \cup Q$ is

$$A_\cup(r) = 1 - r^2.$$

As we increase the parameter r towards 1, the sizes of R and S decrease, but at the same time α^2 increases, making good mixing more likely. On the other hand, choosing a small value for r in order to maximize the size of R and S results in a small value of α^2, and so mixing is more likely to fail in the sense that (9.1) and (9.2) may not be satisfied. This information is illustrated quantitatively in Figure 9.1.

The radius parameter r is varied on the horizontal axis from 0 to 1. The areas of the union and intersection of the annuli, $A_\cup$ and $A_\cap$ respectively, are drawn using the left-hand vertical axis. Using the right-hand vertical axis we plot the inequalities (9.1) and (9.2). We mark horizontal lines to show the conditions on the strengths of the twists to guarantee the existence of an ergodic partition, and for the Bernoulli property on the whole region (in the case $k \geq 2$). Also using the right-hand vertical axis we draw the value of α^2 for $k = 1$ (dashed line) and $k = 2$ (dotted line). So considering the line for single-twists, we see that choosing $r < 0.5$ may result in large areas for R and S, but we do not guarantee even an ergodic partition, and so islands may occur (see Section 5.3.4). For $r \in [0.5, 0.75]$ we have sufficient strength of twist to guarantee an ergodic partition, but not to guarantee that the partition only contains one component.

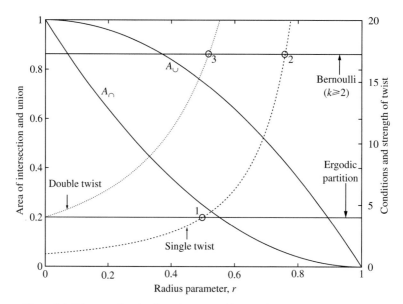

Figure 9.1 Diagram showing how the areas of intersection and union of P and Q depend on the radius parameter r for a symmetric counter-rotating TLTM, together with criteria to guarantee results on ergodicity and the Bernoulli property. The areas $A_\cap(r)$ and $A_\cup(r)$ are shown on the left-hand axis, while the right-hand axis displays the inequalities (9.1) and (9.2).

The point at which a single-twist system gains an ergodic partition is marked 1 on the diagram. For $r \gtrsim 0.75$ (marked 2) we cross the ergodicity condition line, but we cannot conclude that we have the Bernoulli property on the whole region as we only have single-twists. For the double-twist case, it is apparent that the ergodic partition is guaranteed for any $r > 0$ – in other words, as soon as we choose linear functions with double-twists we can be sure no islands will appear for any size of region. However, we have to choose $r \gtrsim 0.55$ (marked 3) to be sure that we have ergodicity on the whole region. In this case, as $k = 2$, we also have verified that the system is Bernoulli.

The behaviour of the symmetric counter-rotating linear toral linked twist map can also be illustrated using the bifurcation diagram in Figure 9.2. This diagram shows the type of behaviour exhibited at different parameter values. On the horizontal axis we again vary the radius parameter r from 0 to 1, and the areas of R and S can be read off the right-hand vertical axis. The left-hand vertical axis gives different wrapping numbers, from single-twists up to $k = 5$. The graph is then shaded according to the different mixing results which are guaranteed. For example, for a single-twist, we see the behaviour depicted in

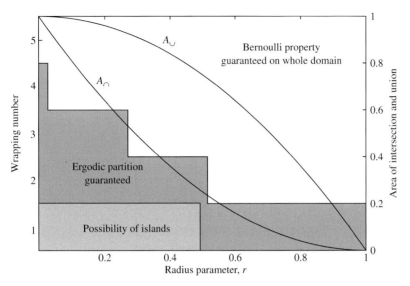

Figure 9.2 Bifurcation diagram displaying the ergodic behaviour of the symmetric counter-rotating TLTM, using r and the wrapping number k as parameters. Superimposed are the areas of R and S on the right-hand axis.

Figure 9.1. For higher wrapping numbers, as might be expected, the Bernoulli property is guaranteed for smaller values of r. In particular, it can be seen that a wrapping number of 5 produces the Bernoulli property for any size of annuli.

9.2.2 Planar linked twist maps

We can perform the corresponding analysis for the planar version of the linked twist maps. Here the geometry of the system makes the calculations a little more involved, but similar results can be achieved. As in Chapter 8 we assume we have two identical annuli centred on $(-1, 0)$ and $(1, 0)$. We fix the outer radii $r_1 = 3.0$, and vary the inner radii r_0 as a parameter between 1.0 and 3.0. We assume as in the previous section that we have linear twist functions on both annuli, and we will investigate both co- and counter-rotating systems.

Area of intersection of a pair of annuli

We first consider the area of intersection of two circles. As in Figure 9.3 take a circle centred on $O_- = (-1, 0)$ with radius r and a circle centred on $O_+ = (1, 0)$ with radius s. We label the two points of intersection of the two circles A (in the

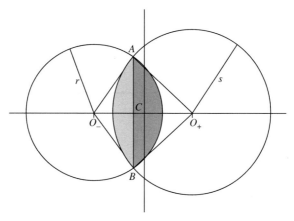

Figure 9.3 The area of intersection between two circles can be computed by subtracting the areas of triangles from the areas of sectors.

upper half-plane) and B (in the lower half-plane). To find the area of intersection of these circles, referring to Figure 9.3, we find the areas of the sectors swept out by the angles $\angle(BO_-A)$ and $\angle(AO_+B)$, and the areas of the triangles O_-AB and O_+AB. The area of the light shaded part of the figure is the area of triangle O_+AB subtracted from the area of the $\angle(BO_+A)$-sector, and similarly for the darker shaded part.

Let C be the point of intersection of the lines AB and O_-O_+. The points A and B lie on both circles, and so satisfy the equations

$$
\begin{aligned}
(x+1)^2 + y^2 &= r^2 \\
(x-1)^2 + y^2 &= s^2.
\end{aligned}
$$

Solving these gives the x-coordinate of the line AB as $x = (r^2 - s^2)/4$ (of course when the circles are of identical size the symmetry of their locations guarantees that AB lies along $x = 0$). This immediately gives the lengths $|O_-C| = 1 + (r^2 - s^2)/4$ and $|CO_+| = 1 - (r^2 - s^2)/4$, and then the Pythagoras theorem gives

$$
|AC| = |CB| = \left(\frac{r^2 + s^2}{2} - \frac{(r^2 - s^2)^2}{16} - 1 \right)^{1/2},
$$

and we have the angles

$$\angle(BO_-A) = 2\cos^{-1}\left(\frac{1 + (r^2 - s^2)/4}{r}\right)$$

$$\angle(AO_+B) = 2\cos^{-1}\left(\frac{1 - (r^2 - s^2)/4}{s}\right).$$

Now since the ratio of the area of a sector to the area of a circle is equal to the ratio of the sector's angle to 2π, we have

$$\text{Area of sector } BO_-A = r^2\cos^{-1}\left(\frac{1 + (r^2 - s^2)/4}{r}\right)$$

$$\text{Area of sector } AO_+B = s^2\cos^{-1}\left(\frac{1 - (r^2 - s^2)/4}{s}\right).$$

The areas of the triangles are straightforwardly

$$\text{Area of triangle } BO_-A = \left(1 + \frac{r^2 - s^2}{4}\right)\left(\frac{r^2 + s^2}{2} - \frac{(r^2 - s^2)^2}{16} - 1\right)^{1/2}$$

$$\text{Area of triangle } AO_+B = \left(1 - \frac{r^2 - s^2}{4}\right)\left(\frac{r^2 + s^2}{2} - \frac{(r^2 - s^2)^2}{16} - 1\right)^{1/2},$$

and so the area of intersection of the two circles is given by

$$
\begin{aligned}
A_\circ(r, s) \quad = \quad & \text{Area of sector } AO_-B + \text{Area of sector } AO_+B \\
& -\text{Area of triangle } BO_-A - \text{Area of triangle } BO_+A \\
= \quad & r^2\cos^{-1}\left(\frac{1 + (r^2 - s^2)/4}{r}\right) + s^2\cos^{-1}\left(\frac{1 - (r^2 - s^2)/4}{s}\right) \\
& -2\left(\frac{r^2 + s^2}{2} - \frac{(r^2 - s^2)^2}{16} - 1\right)^{1/2}.
\end{aligned}
$$

In particular, when the two circles have identical radii we have

$$A_\circ(r, r) = 2r^2\cos^{-1}(1/r) - 2(r^2 - 1)^{1/2}.$$

Now we can compute the area of intersection (consisting of two components) of two annuli of inner and outer radii r_0 and r_1 (for the annulus centred at $(-1, 0)$) and inner and outer radii s_0 and s_1 (for the annulus centred at $(1, 0)$) as

$$A_\cap = A_\circ(r_1, s_1) - A_\circ(r_0, s_1) - A_\circ(r_1, s_0) + A_\circ(r_0, s_0),$$

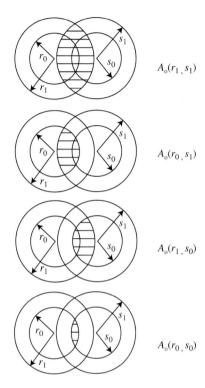

Figure 9.4 Illustration of the four sets (shown hatched) in the formula for $A_\cup$.

and the area of the union of the annuli as

$$A_\cup = \pi(r_1^2 + s_1^2 - r_0^2 - s_0^2) - A_\cap.$$

In Figure 9.4 we illustrate the four sets in the formula for $A_\cup$.

For our choice of identical annuli with $r_1 = s_1 = 3.0$ we have the particular case

$$A_\cap = A_\circ(3,3) - 2A_\circ(r_0, 3) + A_\circ(r_0, r_0)$$

$$A_\cup = \pi(18 - 2r_0^2) - A_\cap.$$

The cotangent of α

Recall that the conditions to guarantee an ergodic partition for both co- and counter-rotating planar linked twist maps depend on the quantity η, which was

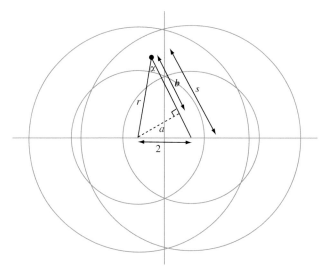

Figure 9.5 The cotangent of the angle α can be computed by finding the lengths of a and b, using the Pythagoras theorem. Then we have $\cot \alpha = b/a$.

given in Chapter 8 as

$$\eta = \sup_{(r,\theta) \in M_1^{-1}(S)} \frac{|\cot \alpha(M_1(r,\theta))|}{r},$$

where α is the angle between lines subtended from a point $(M_1(r,\theta))$ in S to the centres of the annuli. We now give an expression for the cotangent of such an angle α. Referring to Figure 9.5, we note that we require the cotangent of the angle between sides of length r and s of a triangle whose other side is of length 2. Drawing in a perpendicular to the side of length s, we have

$$\cot \alpha = \frac{1}{\tan \alpha} = \frac{b}{a}.$$

To compute a and b we use Pythagoras theorem to give

$$2^2 = (s - b)^2 + a^2$$

$$r^2 = a^2 + b^2,$$

from which we can eliminate a and b to produce

$$a = \sqrt{r^2 - \frac{(r^2 + s^2 - 4)^2}{4s^2}}$$

$$b = \frac{r^2 + s^2 - 4}{2s},$$

and hence

$$\cot \alpha = \frac{r^2 + s^2 - 4}{\sqrt{4r^2 s^2 - (r^2 + s^2 - 4)^2}}.$$

From this expression η can be easily computed numerically by maximizing $|\cot \alpha|/r$ for $r \in [r_0, 3]$ and $s \in [r_0, 3]$.

Criteria for an ergodic partition

The conditions of the theorems in Chapter 8 are based on the strength of the twists. Recalling the definitions from Section 8.2 we note that taking identical linear twists for both annuli, we have the strengths

$$|c| = \frac{2\pi |k|}{3 - r_0},$$

where k is the wrapping number of the twist. Now the conditions in Theorems 8.3.1 and 8.3.2 can be given for the symmetric case as

$$c > 2\eta \qquad\qquad \text{counter-rotating/co-twisting case} \qquad (9.3)$$

$$c^2 - 4c\eta > 4/r_0^2 \qquad \text{co-rotating/counter-twisting case} \qquad (9.4)$$

The counter-rotating planar LTM

Figure 9.6 provides the information for the symmetric counter-rotating/co-twisting planar linked twist map described above, for wrapping numbers $k = 1, \ldots, 4$. We plot the areas of intersection and union, $A_\cup$ and $A_\cap$ as dashed lines using the right-hand vertical axis as a scale. Using the left-hand vertical axis we plot the inequality (9.3). The quantity 2η is plotted for each r_0 as a thick solid line, while values of c for each wrapping number are plotted as dotted lines. Parameter values for which the strengths are larger than 2η give a system with a guaranteed ergodic partition. The relevant crossing points where the theorem applies are marked with circles. Note that unlike the toral case, simply increasing the wrapping number does not automatically guarantee that the condition for the theorem will be fulfilled. However, we recall that these conditions do not appear, based on numerical experiments, to be sharp conditions.

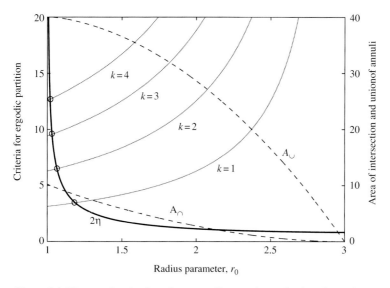

Figure 9.6 Diagram showing how the areas of intersection and union of two planar annuli vary with inner radius r_0, together with the criteria required to deduce that the counter-rotating system has an ergodic partition. The areas $A_\cup$ and $A_\cap$ are plotted as dashed lines using the right-hand vertical axis. The dotted lines give the strength c of the linear twists for $k = 1, \ldots, 4$, and the points where these exceed the quantity 2η are marked with circles.

The co-rotating planar LTM

Figure 9.7 is the corresponding diagram for the symmetric co-rotating/counter-twisting linked twist map. As above, the areas $A_\cup$ and $A_\cap$ are plotted as dashed lines on the right-hand scale, and the inequality (9.4) uses the left-hand axis. The thick solid line gives the quantity $4/r_0^2$, and the dotted lines represent $c^2 - 4c\eta$ for wrapping numbers $k = 1, \ldots, 4$. Again the points at which the ergodic partition theorem applies are marked in circles. Note that, as for the toral case, the counter-twisting system is more restrictive than the co-twisting system, in the sense that to guarantee the ergodic partition, the areas of intersection and union are required to be smaller.

9.3 Breakdown of transversality: effect and mechanisms

As discussed in detail previously, and in for example Ottino (1989a), the fundamental principle behind chaotic mixing in fluids is that streamlines should cross. Moreover, it is also crucial for many mathematical results that they should

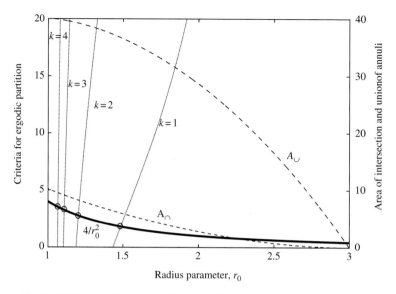

Figure 9.7 Diagram showing how the areas of intersection and union of two planar annuli vary with inner radius r_0, together with the criteria required to deduce that the co-rotating system has an ergodic partition. The areas $A_\cup$ and $A_\cap$ are plotted as dashed lines using the right-hand vertical axis. The dotted lines give the quantity $c^2 - 4c\eta$ of the linear twists for $k = 1, \ldots, 4$, and the points where these exceed the quantity $4/r_0^2$ are marked with circles.

cross *transversely*. This is because, roughly speaking, we would like fluid blobs to be pulled and stretched first in one direction, and then in a different direction. Equally roughly, the more different these directions, the more mixed a system is likely to become. These ideas can be given some rigour by considering some situations for which the theorems of Chapter 8 cannot be applied.

The linked twist map on the torus has a very simple form, which we exploited in Chapter 6 to prove results on the expansion and contraction of tangent vectors with relative ease. This form is due to the fact that the two twist maps not only act in transverse directions everywhere, but actually lie at right angles to each other, so that they can be expressed naturally in a single Cartesian coordinate system. Lack of transversality in the crossing of streamlines (stable and unstable manifolds) was not a concern. On the plane we had to employ coordinate transformations to express the two twist maps in a common coordinate system. In this situation, streamlines did not always cross at right angles, but were guaranteed to be transverse by the construction of the annuli. Recall that two annuli R_1 and R_2 were centred at $(-1, 0)$ and $(1, 0)$ respectively and each given inner

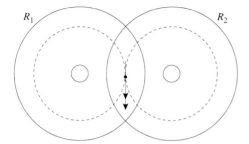 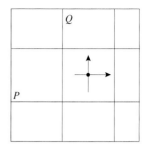

Figure 9.8 How transversality can fail in a planar linked twist map. On the left we show a pair of annuli (marked with solid lines) for which the choice of centres and inner and outer radii give an intersection consisting of a single component. The point marked lies at the point at which the two dotted circles meet tangentially, and so (in the counter-rotating case) the shears from each twist map act in the same direction. By contrast, on the right, in the toral linked twist map case, the directions of shears are at right angles at every point in the domain.

radius $r_0 > 1$ and outer radius $r_1 < 2 + r_0$. This guaranteed that each circle in R_1 intersected every circle of R_2 transversely. It can be easily seen that if we allow the inner radius to be $r_0 < 1$, then (if $r_1 > 1$) $R_1 \cap R_2$ consists of only a single component. The same effect might be achieved by moving the centres of the annuli further apart. In this situation there are now circles of R_1 which intersect a circle of R_2, but tangentially rather than transversely. Alternatively one could create a lack of transversality by allowing $r_1 > 2 + r_0$, or by moving the centres closer together, which again produces a single component intersection. See Figure 9.8.

To understand how and why a lack of transversality affects the mixing properties of a system, consider first the extreme situation of a pair of identical annuli with the same centres. In this instance it is clear to see that the counter-rotating case produces a system for which every point is a period two periodic point (each point is rotated under F_1 and then returned under F_2^{-1}), while the co-rotating case produces another twist map (each point is rotated on the same circle under both F_1 and F_2). See Figure 9.9. Of course, a system in which every point is periodic with period two has no mixing properties at all, while a twist map on an annulus is exactly analogous to a twist map on a torus, discussed in Examples 5.3.2, 5.3.5 and 5.3.7. Recall that in this example taking a vector in the direction of the shear gave no expansion at all, and other vectors resulted in expansion, but at a rate slow enough that the Lyapunov exponent was zero. This results in a system which can only be decomposed into an uncountable number of ergodic components, in the present case one for each circle $r = $ constant in the annulus.

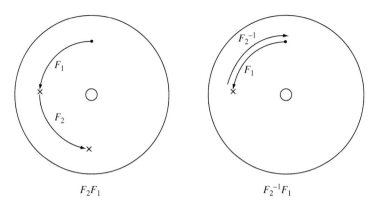

$$F_2F_1 \qquad\qquad F_2^{-1}F_1$$

Figure 9.9 Illustration of the effect of creating a planar linked twist map from two coincident annuli. On the left two co-rotating twist maps F_1 and F_2 results in another twist map F_2F_1, while on the right two counter-rotating twist maps F_1 and F_2^{-1} results in a periodic map of period 2.

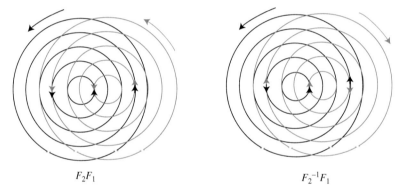

$$F_2F_1 \qquad\qquad F_2^{-1}F_1$$

Figure 9.10 Transverse and tangential intersections for a pair of intersecting annuli. The figure on the left shows the co-rotating case, and the figure on the right the counter-rotating case.

In the more general case in which $R_1 \cap R_2$ consists of a single component we have more complicated dynamics, but retain elements of the non-mixing behaviour described above. In Figure 9.10 we show two such systems, and illustrate some arrows depicting the directions of vectors tangent to the shears. It can be seen that both the co- and counter-rotating versions possess points where such vectors point in the same direction, or in the opposite direction. However, there are still many points at which streamlines cross transversely, indicating that some mixing behaviour might be expected. We shortly give numerical examples illustrating this issue, but first seek to formalize these heuristic ideas.

Loss of transversality strikes at the very heart of hyperbolicity. Recall that for complete hyperbolicity we require a splitting of tangent space into subspaces E^u and E^s. These contain tangent vectors which undergo exponential expansion and contraction (respectively) under iteration of the map. The subspace E^0, which contains vectors which may expand or contract at a slower rate, is required to be empty. Pesin theory equates complete nonuniform hyperbolicity with points giving rise to non-zero Lyapunov exponents for *every* vector in tangent space, and partial nonuniform hyperbolicity with points giving rise to non-zero Lyapunov exponents for only *some* vectors in tangent space.

Previously, in Chapter 8 we noted in Lemma 8.3.8 that the twist map F_1 increased the norm of all vectors in an appropriate cone, *except* for the vector which lay in the direction of the shear; namely $v = (0, \beta_2)$. However, Lemma 8.3.10 provided the result that this vector was mapped under a coordinate transformation into a vector which was expanded by the second twist map F_2 (or F_2^{-1}). The proof of that lemma relied on the fact that the angle α (the angle subtended from any point in the intersection to each of the annuli centres) was not equal to zero or π at each point. Clearly, in the situation in which we have a single intersection region, this proof does not hold. That is, a point lying in the intersection on the $y = 0$ axis in Cartesian coordinates has $\alpha = \pi$ (assuming the centres of the annuli are also on the $y = 0$ axis). This means that taking such a point, a tangent vector of the form $v = (0, \beta_2)$ is expanded by neither F_1 nor F_2. Suppose a trajectory consists of only such points. Then that trajectory has a zero Lyapunov exponent for this choice of v, and the system has only partial hyperbolicity. Thus we cannot deduce a decomposition into countably many ergodic components. Of course, such a trajectory may seem, and indeed be, unlikely. However, if the goal is to rigorously prove nonuniform complete hyperbolicity it cannot be ignored.

Moreover, the conditions in the theorems of Chapter 8, as discussed in the previous section, rely on quantities related to the strength of the twists being greater than some function of α. These conditions are less likely to be fulfilled the greater the cotangent of α.

In many examples of fluid flows designed to promote mixing the effect of breakdown of transversal crossing of streamlines is present. In the following subsections we consider different situations where this can occur.

9.3.1 Separatrices

A situation in which we naturally see a loss of transversality occurs when there is a *separatrix* in the flow. A separatrix in this context could be described as a streamline which divides two distinct regimes of behaviour. We have seen

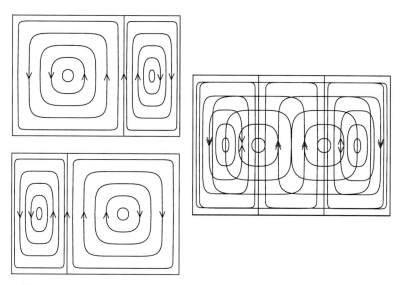

Figure 9.11   An idealized model of streamlines in a blinking flow with separatrices. The diagrams on the left give two streamline patterns, and on the right is the superposition of the two.

examples of this in Chapter 2 in mixers such as the electro-osmotic driven micromixer (Qian & Bau (2002)) shown in Figure 1.8; the 'herringbone' mixer (Stroock *et al.* (2002)) in Figure 2.11; and the partitioned pipe mixer (Khakhar *et al.* (1987)) in Figure 2.13. Figure 9.11 illustrates an idealized set of stream-lines in a system with a separatrix. If the patterns on the left represent the streamlines of a two-dimensional blinking flow (or the cross-sections of a duct flow), then the superposition on the right gives the streamline crossing picture, with many non-transverse intersections.

9.3.2 Planar linked twist maps with a single intersection component

Recall the blinking flow described in Section 1.5.1. In the linked twist map framework, this corresponds to a pair of planar annuli with inner radius zero, corresponding to the fact that the vortices act on all points, however close to the vortex.

The results for linked twist maps in the plane required that the two annuli intersect in two disjoint components. Here we consider what may happen when this condition is violated.

We denote the centre of the annuli R_i by (x_i, y_i) (in Cartesian coordinates). All of the previous theory relates to the case $(x_1, y_1) = (-1, 0)$, $(x_2, y_2) = (1, 0)$. Here in Figure 9.12 we fix the inner and outer radii $r_0 = 0.2, r_1 = 3.0$, and fix the strength of the (linear) twist $\alpha = 1.0$. We fix the centre of R_2 at $(x_2, y_2) = (0, 0)$, and vary the centre of R_1. The left-hand column of the figure shows the evolution of two initial blobs marked in red and blue in the top figure. The figures below show the images of the blobs under 1, 3, 6 and 10 iterations of $\bar{T}$ respectively. In this left-hand column we set $(x_1, y_1) = (-0.8, 0)$. Clearly after ten iterations the blobs are poorly mixed. In the centre column we show the evolution of the same initial blobs for the case $(x_1, y_1) = (-1, 0)$. With the annuli separated slightly (and hence a slightly smaller region of intersection) the mixing process appears more effective. Finally the right-hand column shows the same experiment with the annuli separated slightly more, with $(x_1, y_1) = (-1.4, 0)$. After ten iterations the blobs are well mixed.

Heuristically, this behaviour may be expected from the arguments above. Mixing through chaotic advection stems from the crossing of streamlines, and in general, the more transverse the crossing, the better. When the centres of the annuli are separated only a small amount, the streamlines generated by the twists in each annulus at each point run in roughly the same direction. The angles between them become larger the more separated the annuli become. At the same time, separating the annuli more means shrinking the size of the intersection. This is what brings in a design issue here – how to maximize the size of the intersection whilst achieving the desired quality of mixing.

9.3.3 More than two annuli

In the toral linked twist map framework, it is straightforward to extend results to any number of linked horizontal and vertical annuli (Devaney (1980)). The geometric complications of linked twist maps on the plane makes the corresponding theory more difficult. However, it is a natural question to ask how the introduction of another annulus (or more) may affect the quality of mixing.

Figure 9.13 illustrates how poor mixing may be improved by the introduction of a third annulus. The linked twist map for three annuli in the plane is constructed in an obvious way. Thus for $i = 1, 2, 3$ we have three annuli R_i with centres (x_i, y_i), each with a twist map T_i, where as usual the nature of a twist map T_i is to apply a twist to points in R_i, and leave unchanged those outside. We form the linked twist map $T = T_3 T_2 T_1$ for the following numerics. In each of the plots in Figure 9.13 we fix the inner and outer radii of all three annuli as $r_0 = 0.2$, $r_1 = 3.0$, and fix the strength of each (linear) twist as $\alpha = 1$. The left-hand column shows the evolution of two blobs for two annuli with centres

Figure 9.12 Numerical results for three experiments for planar linked twist maps
for which the intersection of the annuli has a single component. In each case we
have linear single-twists (strength $\alpha = 1$), and each annulus has radii $r_0 = 0.2$,
$r_1 = 3.0$. In each case R_2 is centred at the origin $(0,0)$, and we vary the centre of R_1
in each of the three columns of the figure. Each column shows the evolution of two
initial blobs shown in the top figure under 1, 3, 6 and 10 iterations of the linked twist
map. In the left-hand column $(x_1, y_1) = (-0.8, 0)$ results in poor mixing; in the
middle column the mixing is improved for $(x_1, y_1) = (-1.0, 0)$; in the right-hand
column for $(x_1, y_1) = (-1.4, 0)$ we have more much effective mixing.

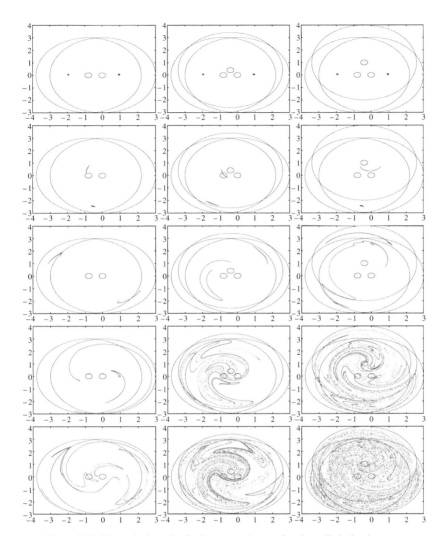

Figure 9.13 Numerical results for three experiments for planar linked twist maps with three annuli. The first column is as the first column in Figure 9.12; that is, $\alpha = 1$, $r_0 = 0.2$, $r = 3.0$ and offset -0.8. We show the images of initial blobs under 1, 3, 6 and 10 iterations. The second column introduces a third annulus at $(-0.4, 0.4)$, and the third column places the extra annulus at $(-0.4, 1.0)$.

$(x_1, y_1) = (-0.8, 0)$, $(x_2, y_2) = (0, 0)$. The initial blobs in the top figure are shown under 1, 3, 6 and 10 iterations of the linked twist map. (This column is the same as the left-hand column of Figure 9.12.) In the middle column we show the evolution of the same blobs with the introduction of a third annulus, with centre $(x_3, y_3) = (-0.4, 0.4)$. The right-hand column shows the same experiment with R_3 centred at $(x_3, y_3) = (-0.4, 1)$.

Just as in the previous example, the change in the quality of mixing can be understood heuristically by considering the angle between streamlines. The introduction of a third annulus only becomes beneficial to mixing when it is placed sufficiently far from the others to create enough transversality between streamlines.

9.4 Monotonicity of the twist functions

Another key component behind the mathematical results for linked twist maps is the requirement that the twist functions be *monotonic*. That is, the twist functions must not have any turning points; if they have regions with positive derivative, there must not be any regions with negative derivative. Recall that points of inflexion can be dealt with using the technique of Section 6.3. The reason monotonicity is so vital is that we cannot run the risk of any stretching being undone, or reversed, by contraction at a later iterate. In other words, we wish to prevent mixing for some iterates and then 'unmixing' for later iterates.

Suppose we have a linked twist map H made up of twist maps F and G, where F has a nonmonotonic twist function $f(y)$, while G has a monotonic twist function $g(x)$ such that $g'(x) > 0$ for each x. Consider an orbit beginning at a point $z_0 = (x_0, y_0)$ for which $f'(y_0) = \alpha_1 > 0$. Then an initial tangent vector $v = (v_1 > 0, v_2 > 0)$ at z_0 is expanded under DF, since

$$\|DFv\| = \left\| \begin{pmatrix} 1 & \alpha_1 \\ 0 & 1 \end{pmatrix} \begin{pmatrix} v_1 \\ v_2 \end{pmatrix} \right\| = \left\| \begin{pmatrix} v_1 + \alpha_1 v_2 \\ v_2 \end{pmatrix} \right\|,$$

which gives $\|DFv\| > \|v\|$ as $\alpha_1 > 0$. Since G is such that $g'(x) > 0$ for each x, the vector $DHv = DGDFv = \tilde{v} = (\tilde{v}_1, \tilde{v}_2)$ remains in the positive quadrant of tangent space. Now suppose that $GF(x, y) = (x_1, y_1) = z_1$ is such that $f'(y_1) = \alpha_2 < 0$. Then

$$\|DF\tilde{v}\| = \left\| \begin{pmatrix} 1 & \alpha_2 \\ 0 & 1 \end{pmatrix} \begin{pmatrix} \tilde{v}_1 \\ \tilde{v}_2 \end{pmatrix} \right\| = \left\| \begin{pmatrix} \tilde{v}_1 + \alpha_2 \tilde{v}_2 \\ \tilde{v}_2 \end{pmatrix} \right\|,$$

but since $\alpha_2 < 0$ it is now no longer true that $\|DF\tilde{v}\| > \|\tilde{v}\|$. This is illustrated in Figure 9.14. Moreover it is straightforward to see that if G is also nonmonotonic then it is even easier for tangent vectors to be expanded or contracted without regularity at each iteration. This illustrates the problem with constructing invariant expanding (or contracting) cones for nonmonotonic linked twist maps. While it is possible that a trajectory falls into the appropriate region of

Figure 9.14 Sketch to illustrate how lack of monotonicity may cause expansion of tangent vectors to fail. The twist functions f and g on the annuli P and Q are shown as dotted lines – we take f to be nonmonotonic and g to be linear. In the first figure the initial point z_0 (in grey) is mapped under F while the initial tangent vector v (also in grey) is expanded under DF. The second figure shows $F(z_0)$ and DFv (in grey) and their images z_1 and $\tilde{v}$ under G and DG respectively. The tangent vector is again expanded under the Jacobian map. Finally the third figure shows the next iteration of F and DF, which maps z_1 and $\tilde{v}$ into the black point and vector. Under this iteration the tangent vector is contracted by the Jacobian map.

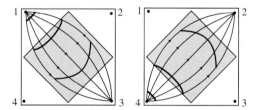

Figure 9.15 The nonmonotonicity of a pulsed source-sink system. A set of tracer particles leaving a source forms the velocity profile of the twist function.

a twist function's domain to produce expanding behaviour at every iterate (and hence a positive Lyapunov exponent), without a priori knowledge about the behaviour of the trajectory it is impossible to deduce that this will occur for typical trajectories.

In many examples however, it may be that nonmonotonic twist functions are typical. Consider for example the DNA pulsed source-sink pair mixer described in Section 1.5.8. Recall that we have a twist map structure on a source-sink system because a fluid particle travels quickest from source to sink along the straight line connecting the source-sink pair, while particles travelling along a curved line on either side of this shortest path travel more slowly. Hence the twist condition breaks down along the straight line path, as shown in Figure 9.15. This lack of monotonicity means that the Bernoulli property cannot be guaranteed in the whole domain, and that islands of unmixed fluid are possible.

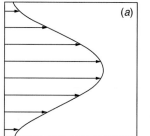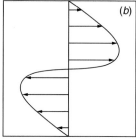

Figure 9.16 Twist functions for the egg beater flow model, taken from Franjione & Ottino (1992). The egg beater is modelled as a linked twist map in the form of a nonmonotonic Arnold Cat Map.

Nonmonotonic twist maps have been studied as systems in their own right (for example Howard & Humpherys (1995) studied perturbations of nonmonotonic twist maps of the type described in this book), and can produce rich and complex behaviour. In the context of linked twist maps as paradigms for mixers they are interesting because prescribed boundary conditions may make them inevitable components of the mixer in question. For example, consider the egg beater flow (Franjione & Ottino (1992)) discussed in Section 1.5.6. This is modelled as a composition of two twist maps. It could be viewed as a generalized version of the Arnold Cat Map. Recall that the Cat Map of Example 5.2.4 is equivalent to the composition of two linear, co-rotating twist maps on the whole torus T^2. The egg beater flow allows the twist functions to be more general, with the restriction, for physical reasons, that they have non slip boundary conditions. Two typical twist functions, or velocity profiles, are shown in Figure 9.16, taken from Franjione & Ottino (1992). This system can be made to exhibit a wide variety of behaviours, from completely regular to completely chaotic, depending on the details of the system. Such aspects, for example the relative direction, and the even- and odd-ness, of the shears have been studied in detail in Franjione & Ottino (1992), Ottino *et al.* (1992), Ottino (1989b). In the following section we give a sample of the types of behaviour possible in nonmonotonic toral linked twist maps.

9.4.1 Lack of monotonicity in toral linked twist maps

As stated earlier, a crucial element which allows the construction of all the proofs in previous chapters is that twist functions are monotonic. This means there are no turning points in the twist functions, and so we have a derivative which is everywhere positive, or negative, but never a mixture of the two. A different approach may be required to prove similar results for systems in which the twist functions are nonmonotonic. Here we give numerical evidence that again an important factor governing the ergodic properties of such a map is a notion of the 'strength' of a twist. Previously we have defined the strength of a twist to be the shallowest slope of the twist function. Clearly if the twist function has a turning point, the strength will be equal to zero. Instead, for nonmonotonic functions, one might define the strength to be equal to the steepest slope, or perhaps the average derivative.

We take twist functions to be the nonmonotonic functions $f(y) = a(y - y_0)(y_1 - y)$ and $g(x) = b(x - x_0)(x_1 - x)$. Note that $f(y)$ has a turning point when $f'(y) = 0$, which occurs at $y = (y_0 + y_1)/2$; that is, at the midpoint of the width of the annulus P. Similarly the turning point for $g(x)$ occurs at the midpoint of Q. In Figure 9.17 we take $x_0 = y_0 = 0$ and $x_1 = y_1 = 1$ so that the annuli P and Q coincide with the whole torus T^2, and we study the symmetric case $a = b$. Similar pictures can be obtained taking P and Q smaller than T^2, but as usual decreasing the size of the intersection makes good mixing more likely. In this example the twist functions become $f(y) = ay(1 - y)$ and similarly for $g(x)$, which many readers will recognize as the logistic map with parameter a. This system could be viewed as a pair of alternating, or blinking, logistic maps on the torus.

We investigate the ergodic properties of the map for different values of the parameter a. Whilst the strength of the twist as we have defined it in Chapter 6 is clearly zero, due to the turning point in the twist functions, increasing a has the effect of increasing the maximum slope of f and g, and also the average of the absolute value of the derivative. Figure 9.17 shows the behaviour of this system for four different values of a. On the left we show the twist functions $f(y)$ and $g(x)$, while on the right we show a selection of trajectories. The island structure is clearly visible in each of the first three figures, for $a = 1.0$, 1.3, 1.8. In the final figure, for $a = 2.4$, a single initial condition give a trajectory apparently ergodic on the entire domain.

9.4.2 Non-slip boundary conditions with breakdown of monotonicity

Suppose we wish to impose non-slip boundary conditions at both inner and outer radii of a planar annulus. This may lead to a twist function which is nonmonotonic (ignoring the trivial case in which there is no twist anywhere). Allowing the twist function to be nonmonotonic complicates the dynamics seriously, as there are now regions of expansion and contraction in the same annulus, and

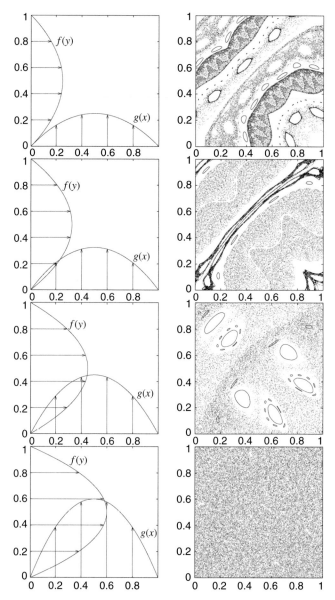

Figure 9.17 Numerical results for toral linked twist maps with nonmonotonic twist functions. Figures on the left show the twist functions, and figures on the right show a selection of trajectories. The four parameter values are $a = 1.0, 1.3, 1.8, 2.4$.

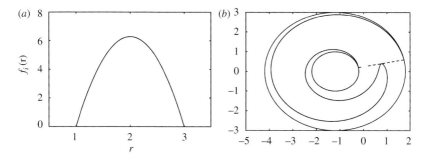

Figure 9.18 Figure (a) shows the twist function f_i for $\alpha = 1$, which is zero on inner and outer radii ($r_0 = 1.0, r_1 = 3.0$), and reaches a maximum of 2π midway between. Figure (b) shows an initial line (dotted) and its image (solid) under one iteration of the twist map. Since $\alpha = 1$ the point on the initial line midway between inner and outer radii makes exactly one complete turn of the circle.

points can be taken either clockwise or anti-clockwise depending on their position in the annulus. Some of the difficulties in introducing nonmonotonicity are illustrated in this section, but we recall that even in the simpler case of toral linked twist maps, the theory requires monotonicity in the twist functions.

Here we fix the twist functions to be

$$f_i(r) = \frac{8\pi \alpha (r - r_0)(r_1 - r)}{(r_1 - r_0)^2}$$

so that $f_i(r_0) = 0, f_i(r_1) = 0$ and the twist reaches a maximum halfway between inner and outer radii (that is, $f'((r_1 + r_0)/2) = 0$). At this point we have $f((r_1 + r_0)/2) = 2\pi\alpha$ so that $\alpha = 1$ corresponds to the point of greatest twist making a complete circuit of the annuli. Figure 9.18 shows the twist functions $f_i(r)$ and the image of an initial line in the annulus R_1 under a single iteration of F_i.

In Figure 9.19 we illustrate drastically different behaviour for planar linked twist maps for which the twist condition breaks down. In each plot we fix the outer radii $r_1 = 3.0$, the strength $\alpha = 1$, and the centres at $(x_1, y_1) = (-0.7, 0)$ and $(x_2, y_2) = (0, 0)$. The only parameter we vary is r_0. In the left-hand column the inner radii are set to $r_0 = 0.2$. As before the top picture shows two initial blobs and the figures beneath show the images of the blobs under 1, 3, 6 and 10 iterations of the linked twist map. Whilst the blobs are spread over a large region of the domain, it is clear that almost no mixing has taken place. Indeed this appears more like a pair of distinct ergodic components than a single mixed region. There are also isolated islands above and below the centres of the annuli.

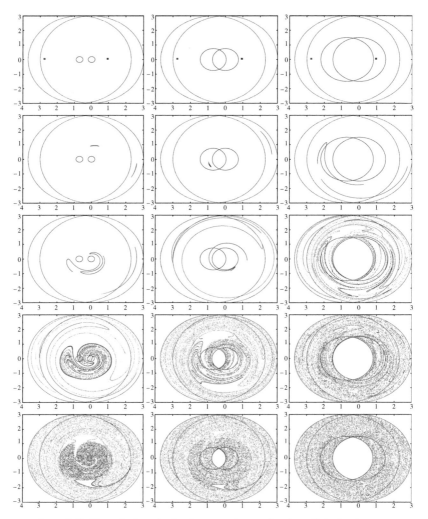

Figure 9.19 Numerical results for three experiments for planar linked twist maps in which the twist condition breaks down. Each annulus has $r_1 = 3.0$, and the centres are at $(x_1, y_1) = (-0.7, 0)$ and $(x_2, y_2) = (0, 0)$. We take $\alpha = 1$ and vary r_0. The three columns are for $r_0 = 0.2, 0.75, 1.5$ respectively, and the initial blobs in the top figures are shown under 1, 3, 6 and 10 iterations. Mixing seems to improve on increasing the inner radii.

Increasing the inner radii in the middle column of Figure 9.19 to $r_0 = 0.75$ we see the mixing begin to improve, although the islands still persist, and for much large inner radii ($r_0 = 1.5$) in the right-hand column the system appears to mix well.

9.4.3 Non-slip boundary conditions

All the linked twist maps discussed above have the property that while points on the outer radii are fixed (modulo the circumference), the outer radii themselves are rotated by an integer multiple of the circumference. Many applications bring in non-slip boundary conditions, which could result in outer (and inner) radii being fixed, so that points on the circumference do not move at all. In order to retain monotonicity in the twist functions f_i we set the inner radii $r_0 = 0$, and fix

$$f_i(r) = 2\pi\alpha \left(1 - \frac{r}{r_1} \right)$$

so that there is no twist on the outer radii.

In Figure 9.20 we fix the centres $(x_1, y_1) = (-1.2, 0)$, $(x_2, y_2) = (0, 0)$, and the outer radii $r_1 = 3.0$. In this experiment we use linear twist functions and vary the strength α of the twist. The left-hand column shows the behaviour of the system for $\alpha = 0.5$. The top figure demonstrates the action of the twist in one annulus R_1. The image of the initial red line after one iteration of F_1 is shown in blue. Beneath that we show two initial blobs, followed by their evolution under 3, 6 and 10 iterations of $\bar{T}$. Perhaps unsurprisingly, with this relatively weak twist, there is little mixing. The middle column shows the same numerics but for $\alpha = 1.0$, and the right-hand column has $\alpha = 2.0$. Of course, the stronger the twist, the greater the quality of mixing. However, in applications, a stronger twist would probably imply a greater input of energy, and so the design principle would involve maximizing quality of mixing whilst minimizing the strength of the twist.

9.5 Final remarks

The coverage of this book is manifestly focused: an examination of the Linked Twist Map (LTM) as it applies to mixing. It is one issue meeting many. Mixing problems are varied in scope and it is unreasonable to expect that an LTM-focused perspective will answer all questions or even have something valuable to contribute to all possible mixing – even if idealized – examples. In fact, in the preceding paragraphs we highlighted a few of the mathematical issues that need to be addressed to increase the scope of the LTM framework. All these caveats notwithstanding, it is clear nevertheless that the LTM captures and formalizes the central heuristic of mixing in 2D flow: the crossing of streamlines. We believe that the potential for the application of the LTM formalism to mixing is high, even higher than the preceding chapters may indicate. There are however

Figure 9.20 Numerical results for three experiments for planar linked twist maps with non-slip boundary conditions. Each annulus has $r_0 = 0.0$, $r_1 = 3.0$ with centres at $(x_1, y_1) = (-1.2, 0)$ and $(x_2, y_2) = (0, 0)$. The top figures show the effect (blue) of applying a twist to an initial line (red) in the cases $\alpha = 0.5$ on the left, $\alpha = 1.0$ in the middle, and $\alpha = 2.0$ on the right. Beneath these are, for these values of α, two initial blobs and their images under 3, 6 and 10 iterations of the linked twist map. Predictably, increasing the strength of twist increases the quality of mixing.

several knowledge gaps and we have described a few of the mathematical issues that need to be addressed to close them. This is a fertile area for research and new mathematical directions, where the methods of ergodic theory and Pesin theory can be used to guide quantitative modelling and analysis. We close by briefly mentioning a few other topics and areas that should provide fruitful avenues for investigation.

Pseudo-Anosov maps and topological mixing In recent years techniques from the study of the topology of maps of surfaces (Boyland (1994)) have been applied to mixing in macroscopic flows by Boyland *et al.* (2000). These pseudo-Anosov maps also have the Bernoulli property. A direct connection with LTMs can be made through the blinking vortex flow (Kin & Sakajo (2005)). While the mathematical language may appear formidable (and we will not go into it here), once it has been translated into the context of a mixing problem it becomes almost intuitive. This is a line of research that is being pursued vigorously (see, e.g., Thiffeault (2005), Vikhansky (2004), and Finn *et al.* (2003)). MacKay (2001) has written an instructive review of the topic. However, we caution the reader that 'topological notions' are generally of a different character from 'measure-theoretic' notions. For example, general conditions do not appear to be known in the nonuniformly hyperbolic setting for which the chaotic invariant set that follows from the map being pseudo-Anosov has positive measure.

Diffusion Numerous articles have dealt with the effect of molecular diffusion on chaotic flows. A typical way of modelling the effect of molecular diffusion is to add a noise term to the advection equations, see, e.g., Jones (1991), or Camassa & Wiggins (1991). In recent years there has been great progress in the development of the notion of a 'random dynamical system'. Potentially this could be used to gain new insights into the interaction of advection and diffusion when coupled with the ergodic theory framework described in this book. See, e.g., Arnold (1998), Kifer (1988), and Cowieson & Young (2005).

Decay of correlations As mentioned in Chapter 3 the decay of correlations is in a rough sense a measure of the speed of mixing. Baladi (2001) gives a review of the subject. While this is an area that is at the forefront of research in ergodic theory, the systems examined to date appear to have little relevance to fluid mechanical mixing problems. For example, there are no known results for two-dimensional, area-preserving, nonuniformly hyperbolic maps. The linked twist maps could be an ideal candidate for rigorous studies of the decay of correlations for such systems.

Three-dimensional linked twist maps Thus far the linked twist map formalism has only been applied to two-dimensional maps. However, there are obvious generalizations to three dimensions using the action-angle-angle coordinates and action-action-angle coordinates in Mezic & Wiggins (1994). These coordinates are very much the same as the coordinates used in constructing the LTM for the duct flow in Chapter 2. Mullowney *et al.* (2005) have already constructed an analogue of the blinking vortex flow in three dimensions (the 'blinking roll' flow).

We hope that this book serves as an invitation to explore an exciting area of analysis and applications.

References

Adler, R. L. (1998). Symbolic dynamics and Markov partitions. *Bulletin of the American Mathematical Society*, **35**, 1–56.

Alekseev, V. M. (1969). Quasirandom dynamical systems. *Mathematics of the USSR Sbornik*, **7**, 1–43.

Anderson, J. L. & Idol, W. K. (1985). Electroosmosis through pores with nonuniformly charged walls. *Chemical Engineering Communications*, **38**(3–6), 93–106.

Anderson, P. D., Galaktionov, O. S., Peters, G. W. M., van de Vosse, F. N., & Meijer, H. E. H. (2000). Mixing of non-Newtonian fluids in time-periodic cavity flows. *Journal of Non-Newtonian Fluid Mechanics*, **93**, 265–86.

Anosov, D. (1967). Geodesic flows on closed Riemannian manifolds with negative curvature. *Proceedings of the Steklov Institute of Mathematics*, **90**, 1–235 (AMS Translations, 1969).

Anosov, D. & Sinai, Y. G. (1967). Certain smooth ergodic systems. *Russian Mathematical Surveys*, **22**, 103–67.

Aref, H. (1984). Stirring by chaotic advection. *Journal of Fluid Mechanics*, **143**, 1–21.

Arnold, L. (1998). *Random Dynamical Systems*. Springer-Verlag.

Arnold, V. I. (1978). *Mathematical Methods of Classical Mechanics*. Springer-Verlag.

Arnold, V. I. & Avez, A. (1968). *Ergodic Problems of Classical Mechanics*. Addison-Wesley.

Baladi, V. (2001). Decay of correlations. In *Smooth ergodic theory and its applications (Seattle WA, 1999)*, volume 69 of *Proceedings of Symposia in Pure Mathematics*, pages 297–325, American Mathematical Society.

Barreira, L. & Pesin, Y. B., (2002). *Lyapunov Exponents and Smooth Ergodic Theory*, volume 23 of *University Lecture Series*. American Mathematical Society.

Beerens, S. P., Ridderinkhof, H., & Zimmerman, J. T. F. (1994). An analytical study of chaotic stirring in tidal areas. *Chaos, Solitons, and Fractals*, **4**(6), 1011–29.

Beigie, D., Leonard, A., & Wiggins, S. (1994). Invariant manifold templates for chaotic advection. *Chaos, Solitons, and Fractals*, **4**(6), 749–868.

Bertsch, A., Heimgartner, S., Cousseau, P., & Renaud, P. (2001). Static micromixers based on large-scale industrial geometry. *Lab on a Chip*, **1**, 56–60.

Birkhoff, G. D. (1931). Proof of the ergodic theorem. *Proceedings of the Academy of Sciences USA*, **17**, 656–60.

Boffeta, G., Lacorata, G., Redaelli, G., & Vulpiani, A. (2001). Detecting barriers to transport: a review of different techniques. *Physica D*, **159**, 58–70.

Boyland, P. (1994). Topological methods in surface dynamics. *Topology and its Applications*, **58**, 223–98.

Boyland, P., Aref, H., & Stremler, M. A. (2000). Topological fluid mechanics of stirring. *Journal of Fluid Mechanics*, **403**, 277–304.

Brin, M. & Pesin, Y. (1974). Partially hyperbolic dynamical systems. *Mathematics of the USSR-Izvestia*, **8**, 177–218.

Brin, M. & Stuck, G. (2002). *Introduction to Dynamical Systems*. Cambridge University Press.

Bunimovich, L. A., Dani, S. G., Dobrushin, R. L., Jakobson, M. V., Kornfeld, I. P., Maslova, N. B., Pesin, Y. B., Sinai, Y. G., Smillie, J., Sukhov, Y. M., & Vershik, A. M. (2000). *Dynamical Systems, Ergodic Theory and Applications*, volume 100 of *Encyclopædia of Mathematical Sciences*. Second edition. Edited by Ya. G. Sinai, Springer-Verlag.

Burton, R. & Easton, R. (1980). Ergodicity of linked twist mappings. In *Proc. Internat. Conf., Northwestern Univ., Evanston, Ill., 1979*, volume 819 of *Lecture Notes in Math.*, pp. 35–49. Springer-Verlag.

Camassa, R. & Wiggins, S. (1991). Chaotic advection in a Rayleigh–Bénard flow. *Physical Review A*, **43**(2), 774–797.

Chernov, N. & Haskell, C. (1996). Nonuniform hyperbolic K-systems are Bernoulli. *Ergodic Theory and Dynamical Systems*, **16**(1), 19–44.

Chernov, N. & Markarian, R. (2003). *Introduction to the Ergodic Theory of Chaotic Billiards*, Second Edition. IMPA Mathematical Publications.

Chien, W. L., Rising, H., & Ottino, J. M. (1986). Laminar mixing and chaotic mixing in several cavity flows. *Journal of Fluid Mechanics*, **170**, 355–77.

Childress, S. & Gilbert, A. D. (1995). *Stretch, Twist, Fold: The Fast Dynamo*, volume 37 of *Lecture Notes in Physics*. Springer-Verlag.

Conway, J. B. (1990). *A Course in Functional Analysis*, volume 96 of *Graduate Texts in Mathematics*. Springer-Verlag.

Cowieson, W. & Young, L.-S. (2005). SRB measures as zero-noise limits. *Ergodic Theory and Dynamical Systems*, **25**, 1115–38.

Dankner, A. (1978). On Smale's Axiom A dynamical systems. *Annals of Mathematics*, **107**, 517–53.

Devaney, R. L. (1978). Subshifts of finite type in linked twist mappings. *Proceedings of the American Mathematical Society*, **71**(2), 334–8.

Devaney, R. L. (1980). Linked twist mappings are almost Anosov. In *Proc. Internat. Conf., Northwestern Univ., Evanston, Ill., 1979*, volume 819 of *Lecture Notes in Math.*, pages 121–45. Springer-Verlag.

Dugundji, J. (1966). *Topology*. Allyn and Bacon.

Easton, R. (1978). Chain transitivity and the domain of influence of an invariant set. In *Lecture Notes in Mathematics*, volume 668, pp. 95–102. Springer-Verlag.

Fiedor, S. J. & Ottino, J. M. (2005). Mixing and segregation of granular matter: Multi-lobe formation in time-periodic flows. *Journal of Fluid Mechanics*, **533**, 223–36.

Finn, M. D., Cox, S. M., & Byrne, H. M. (2003). Topological chaos in inviscid and viscous mixers. *Journal of Fluid Mechanics*, **493**, 345–61.

Fountain, G. O., Khakhar, D. V., & Ottino, J. M. (1998). Visualization of three-dimensional chaos. *Science*, **281**, 683–6.

Franjione, J. G. & Ottino, J. M. (1992). Symmetry concepts for the geometric analysis of mixing flows. *Philosophical Transactions of the Royal Society of London*, **338**, 301–23.

Galaktionov, O. S., Anderson, R., Peters, G., & Meijer, H. (2003). Analysis and optimization of Kenics static mixers. *International Polymer Processing*, **18**, 138–50.

Guckenheimer, J. & Holmes, P. (1983). *Nonlinear Oscillations, Dynamical Systems and Bifurcations of Vector Fields*. Springer-Verlag.

Halmos, P. R., (1950). *Measure Theory*. The University Series in Higher Mathematics. D. Van Nostrand Company.

Halmos, P. R. (1956). *Lectures on Ergodic Theory*. Chelsea Publishing.

Halmos, P. R. (1974). *Finite Dimensional Vector Spaces*. Second Edition. Springer-Verlag.

Hasselblatt, B. (2002). *Handbook of Dynamical Systems*, volume 1A, chapter Hyperbolic Dynamics. Elsevier.

Hausdorff, F. (1962). *Set Theory*. Chelsea Publishing.

Hill, K. M., Gilchrist, J. F., Ottino, J. M., Khakhar, D. V., & McCarthy, J. J. (1999a). Mixing of granular materials: A test-bed dynamical system for pattern formation. *International Journal of Bifurcations and Chaos*, **9**(8), 1467–84.

Hill, K. M., Khakhar, D. V., Gilchrist, J. F., McCarthy, J. J., & Ottino, J. M. (1999b). Segregation-driven organization in chaotic granular flows. *Proceedings of the National Academy of Science*, **96**(21), 11701–6.

Horner, M., Metcalfe, G., Wiggins, S., & Ottino, J. M. (2002). Transport enhancement mechanisms in open cavities. *Journal of Fluid Mechanics*, **452**, 199–229.

Howard, J. & Humpherys, J. (1995). Nonmonotonic twist maps. *Physica D*, **80**, 256–76.

Hunt, T. J. & Mackay, R. S. (2003). Anosov parameter values for the triple linkage and a physical system with a uniformly chaotic attractor. *Nonlinearity*, **16**, 1499–510.

Jain, N., Ottino, J. M., & Lueptow, R. L. (2005). Combined size and density segregation and mixing in noncircular tumblers. *Physical Review E.*, **71**, 051301–1:051031–10.

Jana, S. C., Tjahjadi, M., & Ottino, J. M. (1994a). Chaotic mixing of viscous fluids by periodic changes in geometry: Baffled-cavity flow. *American Institute of Chemical Engineers Journal*, **40**, 1769–81.

Jana, S. C., Metcalfe, G., & Ottino, J. M. (1994b). Experimental and computational studies of mixing in complex Stokes flows: The vortex mixing flow and multicellular cavity flows. *Journal of Fluid Mechanics*, **269**, 199–246.

Jones, S. W. (1991). The enhancement of mixing by chaotic advection. *Physics of Fluids*, **3**(5), 1081–86.

Jones, S. W. & Aref, H. (1988). Chaotic advection in pulsed source-sink systems. *Physics of Fluids*, **31**(3), 469–85.

Jones, S. W., Thomas, O. M., & Aref, H. (1989). Chaotic advection by laminar flow in a twisted pipe. *Journal of Fluid Mechanics*, **209**, 335–57.

Katok, A. & Hasselblatt, B. (1995). *Introduction to the Modern Theory of Dynamical Systems*. Cambridge University Press.

Katok, A., & Strelcyn, J.-M. with the collaboration of Ledrappier, F. & Przytycki, F. (1986). *Invariant Manifolds, Entropy and Billards; Smooth Maps with Singularities*, volume 1222 of *Lecture Notes in Mathematics*. Springer-Verlag.

Katznelson, Y. & Weiss, B. (1982). A simple proof of some ergodic theorems. *Israel Journal of Mathematics*, **42**, 291–6.

Keller, G. (1998). *Equilibrium States in Ergodic Theory*. Cambridge University Press.

Kellogg, L. H. (1993). Chaotic mixing in the Earth's mantle. *Advances in Geophysics*, **34**, 1–33.

Khakhar, D. V., Rising, H., & Ottino, J. M. (1986). An analysis of chaotic mixing in two model systems. *Journal of Fluid Mechanics*, **172**, 419–51.

Khakhar, D. V., Franjione, J. G., & Ottino, J. M. (1987). A case study of chaotic mixing in deterministic flows: The partitioned pipe mixer. *Chemical Engineering Sciences*, **42**, 2909–26.

Khakhar, D. V., McCarthy, J. J., Shinbrot, T., & Ottino, J. M. (1997). Transverse flow and mixing of granular materials in a rotating cylinder. *Physics of Fluids*, **9**(1), 31–43.

Khakhar, D. V., McCarthy, J. J., Gilchrist, J. F., & Ottino, J. M. (1999). Chaotic mixing of granular materials in two-dimensional tumbling mixers. *Chaos*, **9**(1), 195–205.

Kifer, Y. (1988). *Random Perturbations of Dynamical Systems*. Birkhauser.

Kin, E. & Sakajo, T. (2005). Efficient topological chaos embedded in the blinking vortex system. *Chaos*, **15**(2), 023111.

Knight, J. (2002). Microfluidics: Honey, I shrunk the lab. *Nature*, **418**, 474–5.

Kryloff, N. & Bogoliouboff, N. (1937). La théorie générale de la mesure dans son application à l'étude des systèmes dynamiques de la méchanique non linéaire. *Annals of Mathematics*, **38**(1), 65–113.

Kusch, H. A. & Ottino, J. M. (1992). Experiments on mixing in continuous chaotic flows. *Journal of Fluid Mechanics*, **236**, 319–48.

Lapeyre, G. (2002). Characterization of finite-time Lyapunov exponents and vectors in two-dimensional turbulence. *Chaos*, **12**(3), 688–98.

Lasota, A. & Mackey, M. C. (1994). *Chaos, Fractals, Noise: Stochastic Aspects of Dynamics*, volume 97 of *Applied Mathematical Sciences*. Second Edition. Springer-Verlag.

Leong, C.-W. & Ottino, J. M. (1989). Experiments on mixing due to chaotic advection in a cavity. *Journal of Fluid Mechanics*, **209**, 463–99.

Lind, D. & Marcus, B. (1995). *Symbolic Dynamics and Coding*. Cambridge University Press.

Maas, L. R. & Doelman, A. (2002). Chaotic tides. *Journal of Physical Oceanography*, **32**, 870–90.

MacKay, R. S. (2001). Complicated dynamics from simple topological hypotheses. *Philosophical Transactions of the Royal Society of London A*, **359**, 1479–96.

Mané, R. (1987). *Ergodic Theory and Differentiable Dynamics*. Springer-Verlag.

McQuain, M. K., Seale, K., Peek, J., Fisher, T. S., Levy, S., Stremler, M. A., & Haselton, F. R. (2004). Chaotic mixer improves microarray hybridization. *Analytical Biochemistry*, **325**, 215–26.

Metcalfe, G., Rudman, M., Brydon, A., Graham, L. J. W., & Hamilton, R. (2006). Composing chaos: An experimental and numerical study of an open duct chaotic flow. *American Institute of Chemical Engineers Journal*, **52**(1), 9–28.

Mezic, I. & Wiggins, S. (1994). On the integrability and perturbation of three-dimensional fluid flows with symmetry. *Journal of Nonlinear Science*, **4**, 157–94.

Moon, F. C. (1987). *Chaotic Vibrations: An Introduction for Applied Scientists and Engineers*. John Wiley and Sons Inc.

Moser, J. (1973). *Stable and Random Motions in Dynamical Systems*. Princeton University Press.

Mullowney, P., Julien, K., & Meiss, J. D. (2005). Blinking rolls: Chaotic advection in a three-dimensional flow with an invariant. *SIAM Journal of Applied Dynamical Systems*, **4**(1), 159–86.

Munkres, J. (1975). *Topology*: A First Course. Prentice-Hall.

Newhouse, S. E. (1988). Entropy and volume. *Ergodic Theory and Dynamical Systems*, **8**, 283–99.

Nicol, M. (1996a). A Bernoulli toral linked twist map without positive Lyapunov exponents. *Proceedings of the American Mathematical Society*, **124**(4), 1253–63.

Nicol, M. (1996b). Stochastic stability of Bernoulli toral linked twist maps of finite and infinite entropy. *Ergodic Theory and Dynamical Systems*, **16**, 493–518.

Ornstein, D. & Weiss, B. (1973). Geodesic flows are Bernoullian. *Israel Journal of Mathematics*, **14**, 184–98.

Ornstein, D. & Weiss, B. (1998). On the Bernoulli nature of systems with some hyperbolic structure. *Ergodic Theory and Dynamical Systems*, **18**, 441–56.

Oselcdec, V. I. (1968). A multiplicative ergodic theorem. Lyapunov characteristic numbers for dynamical systems. *Transactions of the Moscow Mathematical Society*, **19**, 197–231.

Ott, E. (1993). *Chaos in Dynamical Systems*. Cambridge University Press.

Ottino, J. M. (1989a). *The Kinematics of Mixing: Stretching, Chaos, and Transport*. Cambridge University Press. Reprinted 2004.

Ottino, J. M. (1989b). The mixing of fluids. *Scientific American*, **260**, 56–67.

Ottino, J. M. (1990). Mixing, chaotic advection, and turbulence. *Annual Reviews of Fluid Mechanics*, **22**, 207–54.

Ottino, J. M. (1994). Mixing and chemical reactions: A tutorial. *Chemical Engineering Science*, **49**(24A), 4005–27.

Ottino, J. M. & Khakhar, D. V. (2000). Mixing and segregation of granular materials. *Annual Review of Fluid Mechanics*, **32**, 55–91.

Ottino, J. M. & Wiggins, S. (2004). Designing optimal micromixers. *Science*, **305**, 485–6.

Ottino, J. M., Muzzio, F. J., Tjahjadi, M., Franjione, J. G., Jana, S. C., & Kusch, H. A. (1992). Chaos, symmetry, and self-similarity: Exploiting order and disorder in mixing processes. *Science*, **257**, 754–60.

Ottino, J. M., Jana, S. C., & Chakravarthy, V. S. (1994). From Reynolds's stretching and folding to mixing studies using horseshoe maps. *Physics of Fluids*, **6**(2), 685–99.

Parry, W. (1981). *Topics in Ergodic Theory*. Cambridge University Press.

Pasmanter, R. (1988). Dynamical systems, deterministic chaos, and dispersion in shallow tidal flows. In *Physical Processes in Estuaries*, J. Dronkers and W. V. Leussen, editors, pp. 42–52. Springer-Verlag.

Perry, A. D. & Wiggins, S. (1994). KAM tori are very sticky: Rigorous lower bounds on the time to move away from an invariant Lagrangian torus with linear flow. *Physica D*, **71**, 102–21.

Pesin, Y. B. (1977). Characteristic Lyapunov exponents and smooth ergodic theory. *Russian Mathematical Surveys*, **32**, 55–114.

Petersen, K. (1983). *Ergodic Theory*. Cambridge University Press.

Poincaré, H. (1890). Sur le problème, des trois corps et les équations de la dynamique. *Acta Mathematica*, volume 13, pp. 1–270.

Pollicott, M. (1993). *Lectures on Ergodic Theory and Pesin Theory on Compact Manifolds*. Number 180 in London Mathematical Society Lecture Note Series. Cambridge University Press.

Pollicott, M. & Yuri, M. (1998). *Dynamical Systems and Ergodic Theory*. Volume 40 of London Mathematical Society student texts. Cambridge University Press.

Przytycki, F. (1983). Ergodicity of toral linked twist mappings. *Annales Scientifiques de l'Ecole Normale Supérieure (4)*, **16**, 345–54.

Przytycki, F. (1986). Periodic points of linked twist mappings. *Studia Mathematica*, **83**, 1–18.

Qian, S. & Bau, H. H. (2002). A chaotic electroosmotic stirrer. *Analytical Chemistry*, **74**, 3616–25.

Raynal, F., Plaza, F., Beuf, A., Carrière, P., Souteyrand, E., Martin, J.-R., Cloarec, J. P., & Cabrera, M. (2004). Study of a chaotic mixing system for DNA chip hybridization chambers. *Physics of Fluids*, **16**(9), L63–L66.

Ridderinkhof, H. & Zimmerman, J. T. F. (1992). Chaotic stirring in a tidal system. *Science*, **258**(5085), 1107–11.

Robinson, C. (1998). *Dynamical Systems: Stability, Symbolic Dynamics, and Chaos*. CRC Press.

Rom-Kedar, V. & Wiggins, S. (1990). Transport in two-dimensional maps. *Archive for Rational Mechanics and Analysis*, **109**(3), 239–98.

Rudin, W. (1974). *Real and Complex Analysis*. Second Edition. McGraw-Hill.

Samelson, R. S. (1994). Unpublished notes. Woods Hole Oceanographic Institute.

Signell, R. P. & Butman, B. (1992). Modeling tidal exchange and dispersion in Boston harbor. *Journal of Geophysical Research*, **97**(C10), 15,591–606.

Signell, R. P. & Geyer, W. R. (1991). Transient eddy formation around headlands. *Journal of Geophysical Research*, **96**(C2), 2561–75.

Sinai, Y. (1970). Dynamical systems with elastic reflections. *Russian Mathematical Surveys*, **25**, 137–89.

Sinai, Y. (1989). Kolmogorov's work on ergodic theory. *Annals of Probability*, **17**(3), 833–9.

Sinai, Y. (1994). *Topics in Ergodic Theory*. Princeton University Press.

Smale, S. (1967). Differentiable dynamical systems. *Bulletin of the American Mathematical Society*, **73**, 747–817.

Stone, H. A., Stroock, A., & Adjari, A. (2004). Engineering flows in small devices: Microfluidics toward a lab-on-a-chip. *Annual Review of Fluid Mechanics*, **36**, 381–411.

Stroock, A. D., Dertinger, S. K. W., Ajdari, A., Mezic, I., Stone, H. A., & Whitesides, G. M. (2002). Chaotic mixer for microchannels. *Science*, **295**, 647–51.

Thiffeault, J. L. (2004). Stretching and curvature along material lines in chaotic flows. *Physica D*, **198**(3–4), 169–81.

Thiffeault, J. L. (2005). Measuring topological chaos. *Physical Review Letters*, **94**(8), 084502.

Thiffeault, J.-L. & Childress, S. (2003). Chaotic mixing in a torus map. *Chaos*, **13**(2), 502–7.

Vikhansky, A. (2004). Simulation of topological chaos in laminar flows. *Chaos*, **14**(1), 14–22.

Walters, P. (1982). *An Introduction to Ergodic Theory*. Springer-Verlag.

Weiss, B. (1975). The geodesic flow on surfaces of negative curvature. In *Dynamical Systems: Theory and Applications*, volume 38 of *Lecture Notes in Physics*, pp. 224–35. Springer-Verlag.

Whitesides, G. M. & Stroock, A. D. (2001). Flexible methods for microfluidics. *Physics Today*, **54(6)**, 42–8.

Wiggins, S. (1988). *Global Bifurcations and Chaos*. Springer-Verlag.

Wiggins, S. (1992). *Chaotic Transport in Dynamical Systems*. Springer-Verlag.

Wiggins, S. (1999). Chaos in the dynamics generated by sequences of maps, with applications to chaotic advection in flows with aperiodic time dependence. *Zeitschrift für Angewandte Mathematik und Physik*, **50**, 585–616.

Wiggins, S. (2003). *Introduction to Applied Nonlinear Dynamical Systems and Chaos, Second Edition*. Springer-Verlag.

Wiggins, S. & Ottino, J. (2004). Foundations of chaotic mixing. *Philosophical Transactions of Royal Society of London*, **362**(1818), 937–70.

Wojtkowski, M. (1980). Linked twist mappings have the K-property. In *Non-linear Dynamics (Internat. Conf., New York, 1979)*, volume 357 of *Annals of the New York Academy of Sciences*, pp. 65–76.

Wojtkowski, M. (1985). Invariant families of cones and Lyapunov exponents. *Ergodic Theory and Dynamical Systems*, **5**, 145–61.

Yomdin, Y. (1987). Volume growth and entropy. *Israel Journal of Mathematics*, **57**(3), 285–300.

Index

K-property, 93, 158, 160
σ-algebra, 68

action-action-angle coordinates, 269
action-angle, 50, 51
action-angle-angle coordinates, 269
action-angle-axial, 51
adjacency matrix, 109, 117, 123
advection, 1, 3, 4, 9, 15, 17, 24, 28, 43,
    44, 46, 47
Anosov conditions, 154, 157
Anosov diffeomorphism, 127
Arnold Cat Map, 37, 38, 41, 89, 126, 135, 148,
    154, 160, 165, 166, 195
Axiom A, 140

Baker's transformation, 7
Baker's map, 112
Bernoulli property, xix, xx, 46, 60, 94, 102,
    105, 106, 109, 125, 149, 151, 153, 154,
    158, 160, 194, 238–240, 242–245
Birkhoff Ergodic Theorem, 79, 80
blinking cylinder flow, 24
blinking flow, 16
blinking vortex, 24, 113
blinking vortex flow, 14, 31
boundary point, 65
bounded, 67

Cantor set, 66, 98, 119, 124
cavity flow, 17, 113, 114
channel, 4
channel flow, 5

chaos, 9
chaotic, 85
chaotic advection, 35
chaotic dynamics, 126
chaotic mixing, 2, 6
chaotic stirring, 15
closed, 66, 67
closure, 65
co-rotating, 39, 41
co-rotating planar linked twist map,
    221, 233
co-rotating toral linked twist maps,
    168, 242
co-twisting, 40, 41
compact, 66, 67, 98
complement, 66
concentration gradients, 6
cone, 154
cone field, 155, 156
Conley–Moser conditions, 112–114, 118,
    122, 123
connected, 67
converge, 66
counter-rotating, 40, 41, 250
counter-rotating linear toral linked twist
    map, 244
counter-rotating planar linked twist map, 221
counter-rotating toral linked twist maps, 168,
    242
counter-twisting, 41
crossing of streamlines, 13, 241

decay of correlations, 89, 241, 269
dense set, 66

Printed in the United States
by Baker & Taylor Publisher Services